McGRAW-HILL PROBLEMS SERIES IN GEOGRAPHY
Geographic Approaches to Current Problems:
the city, the environment, and regional development

Edward J. Taaffe, Series Editor

Wilfrid Bach
AIR POLLUTION

Richard L. Morrill and Ernest H. Wohlenberg
THE GEOGRAPHY OF POVERTY in the United States

Harold M. Rose
THE BLACK GHETTO: A Spatial Behavioral Perspective

ATMOSPHERIC POLLUTION

WILFRID BACH
Department of Geography
University of Hawaii

McGRAW-HILL BOOK COMPANY
New York St. Louis San Francisco Düsseldorf Johannesburg
Kuala Lumpur London Mexico Montreal New Delhi Panama
Rio de Janiero Singapore Sydney Toronto

Copyright © 1972 by McGraw-Hill, Inc. All rights reserved.
Printed in the United States of America. No part of this publication may be reproduced, stored in a retrieval system, or transmitted, in any form or by any means, electronic, mechanical photocopying, recording, or otherwise, without the prior written permission of the publisher.

07-002819-2

3 4 5 6 7 8 9 0 **MAMM** 7 9 8 7 6 5 4 3 2

Library of Congress Catalog Card Number 78-170869

This book was set in Baskerville by John T. Westlake Publishing Services, and printed and bound by The Maple Press Company. The designer was John T. Westlake Publishing Services. The editor was Janis Yates. Ted Agrillo supervised production.

Cover photographs courtesy of ENVIRONMENTAL PROTECTION AGENCY

To Anneliese, my wife

CONTENTS

Editor's Introduction

Preface

Introduction: Causes of the Present Air Pollution Problems

Chapter 1: Definition, Sources, and Types of Air Pollution

Definition of Air Pollution	3
Sources of Air Pollution	4
Major Types and Concentrations of Air Pollution	8
Summary	13

Chapter 2: Meteorological Aspects of Air Pollution

The Influence of Weather on Air Pollution	15
Solar Energy	16
Stability Conditions	17
Air Flow and Turbulence	19
Scavenging Processes	20
Application of Meteorology to Air Pollution	21
Summary	27
The Influence of Air Pollution on Weather and Climate	28
The Effect of Air Pollution on Local Weather and Climate	28
The Effect of Air Pollution on Global Climate	33
Summary	39

Chapter 3: Health Aspects of Air Pollution

Air Pollution Episodes	43
Major Air Pollution Disasters	43
Worldwide Air Pollution Disasters	48
Lessons from Air Pollution Disasters	50
Summary	51

viii   ATMOSPHERIC POLLUTION

    Recent Research of the Effects of Air Pollution on Health   51
        Medical Research Methods   52
        Major Pollutants Affecting Health   55
        Major Diseases Related to Air Pollution   57
        Summary   61

Chapter 4: Economic Aspects of Air Pollution

    Cost of Air Pollution Damages   64
    The Cost of Cleaner Air   67
    Governmental Cost Sharing in Air Pollution Control   73
    Cost-Benefit Analysis of Air Pollution Control   75
    Incentives for Air Pollution Control   80
    Summary   81

Chapter 5: Technology and Air Pollution

    Radioactive Air Pollution   84
        Basic Terminology   84
        Natural Radioactive Air Pollution   86
        Man-Made Radioactive Air Pollution   86
        Effect of Radioactive Air Pollution on Man   97
        Summary   99

Chapter 6: Measures of Air Pollution Control

    Air Pollution Legislation   102
        Guiding Principles for Air Pollution Legislation   102
        Historical Review of Air Pollution Legislation   103
        The Clean Air Amendments of 1970   104
        Legal Enforcement   113
        Summary   115
    Urban Planning and Air Pollution Control   116
        Green Areas and Air Pollution Control   116
        Urban Renewal and Air Pollution Control   119
        Control of Industrial Pollution   125
        Summary   129

Chapter 7: The Public and Air Pollution Control

    Information Programs   131
    Public Awareness or Apathy   132
    Action Programs   134
    Summary   141

Epilogue: Atmospheric Pollution and our Future   143

EDITOR'S INTRODUCTION

Wilfrid Bach brings an unusual background to the writing of this book on atmospheric pollution. Not only has he been engaged actively for many years in scientific research, but he has been equally active in community decision-making in problems related to air pollution. Professor Bach does not suffer polluters gladly, and has not hestitated to speak out in the courts and the mass media as well as the classroom in instances where there has been clear evidence of pollution.

Despite the author's unequivocal commitment to the cause of cleaner air, he is equally firm and consistent in treating each problem in a rigorous and scientific fashion. Evidence is relentlessly pursued and assembled, often in convenient tabular form. It is then critically evaluated and tested before judgment is passed.

The rigor of Professor Bach's approach to an often popularized and oversimplified topic is also evident in his particularly thorough and scholarly job of documentation. The study of air pollution is a relatively new field, and anyone planning a new course dealing with it will find a strong foundation not only in the text itself, but in the generous array of literature from diverse fields which has been assembled and utilized by the author.

The meteorological aspects of atmospheric pollution discussed in this book are closely linked to the concepts treated in an introductory physical geography course: for example, the relation between lapse rates, atmospheric stability, and plume types; the "heat islands" associated with the presence of cities; and the effects of different sorts of pollution on long-term warming of the atmosphere. The wide ramifications of the subject, however, carry the student well beyond the confines of meteorology, and separate consideration is given to the chemical, medical, legal, economic, and city-planning dimensions of air pollution.

In viewing the first three volumes in the Problems Series as geographic studies, the editor is struck by their diversity of disciplinary perspective. In Richard Morrill's *The Geography of Poverty in the United States* the viewpoint is clearly one of spatial organization; in Harold Rose's *The Black Ghetto* the view is both spatial and regional; in Wilfrid Bach's *Atmospheric Pollution* the view is that of the relation between man and his physical environment. The study of such man-land relationships carries with it stringent requirements in cross-disciplinary knowledge and research—requirements well met by Professor Bach in this survey of a particularly complex and far-reaching aspect of man's impact on his physical environment.

EDWARD J. TAAFFE

PREFACE

The atmosphere has always been polluted to some extent. Since the late 1940s and early 1950s, however, our pollution disasters have become more frequent. Smog episodes are now constantly recurring phenomena in some areas. Air pollution has lately been blanketing larger and larger portions of the globe, and on some occasions the problem has increased to hemispherical proportions.

Awareness of and concern about the continuously deteriorating quality of the air we breathe is manifested in the ever increasing number of books dealing with environmental problems in general. The concerned people who read these books invariably learn about the magnitude of the problem, but find little concrete information that might be effectively used in combating the causes of air pollution.

This text concentrates on specific air pollution problem areas. Chapters are structured to include a descriptive section which introduces the bulk of the information available concerning the specific problem area, followed by an explanatory section which discusses possible solutions.

Work in atmospheric pollution will require specially trained personnel who can respond professionally to the requirements of a problem that spans a wide range of academic disciplines. An interdisciplinary approach is used in this book in the hope of creating the kind of cooperative spirit that must be evidenced if any progress is ever going to be made toward finding an overall solution to the air pollution crisis.

Thorough study of the material presented here should provide the reader with the background information necessary to facilitate effective communication between specialists in the various disciplines related to the field of air pollution. Basically, this book has been designed to be used as the principal text for an introductory course in atmospheric pollution. Since an interdisciplinary text can provide challenging information in areas which lie outside the reader's ken, parts of this book might be used as a supplementary reference for courses taught at any higher level.

Air pollution respects neither political borders nor the artificial divisions between the various academic disciplines. Individuals representing the physical, biological, medical, engineering, and social sciences are all deeply involved in research aimed toward providing realistic solutions to the air pollution problem.

Physicists and research engineers are trying to discover and develop new energy and power sources. Transportation specialists are working on low pollution or pollution-free ground and air transport systems. Environmental engineers are designing, testing, installing, and supervising the performance of the various pollution control devices and techniques that may be used to curb emissions from such sources as industrial stacks and automobiles.

Meteorologists are developing mathematical models that may be used in conjunction with weather information to provide short- and long-range pollution concentration forecasts. The information resulting from these models can also provide useful guidance to urban and regional planners in their selection of industrial sites and to air pollution control officers in the placement of air quality monitoring systems. Climatologists are investigating air pollution trends and looking carefully for signs for local adverse weather or climate modification, or global climatic change that might be attributable to air pollution.

Chemists are identifying the various pollutants in the air and devising methods for their accurate measurement. They are also studying the methods by which pollutants are removed from the air and the processes involved in photochemical smog formation. Biologists are investigating the adverse effects of air pollution on animals and vegetation. Medical scientists are conducting epidemiological and clinical studies of the acute and chronic health problems caused or aggravated by air pollution.

Economists are developing cost-effectiveness models that determine the optimum use of control devices and shed some light on the financial losses suffered by the general public as a result of air pollution damage. Lawyers are scrutinizing air pollution legislation and regulations and are initiating court action when necessary. Social scientists are investigating the state of awareness of the general public concerning air pollution. They are also conducting public opinion surveys to assess the willingness of the general public to support air pollution abatement programs. Government control officials are devising programs for air pollution control including specific actions to be taken during alert and emergency situations.

Private citizens and conservation groups representing the widest possible range of occupations and interests are coming together at public hearings to make their collective desire for breathable air known to the legislators and governmental officials who will be setting regional air quality standards. The fight for clean air is on. Everybody is involved. No one is immune to the ravages of air pollution. Acquiring the tools needed in this battle for survival is truly a matter of self-preservation.

WILFRID BACH

INTRODUCTION
CAUSES OF THE PRESENT AIR POLLUTION PROBLEMS

The earth's atmosphere has served man in two fundamental ways throughout his existence: it has provided him with life-sustaining air to breathe and it has acted as a medium for disposing of his refuse. The atmosphere is equipped to efficiently dispose of reasonable quantities of the wastes associated with human activity. Man's indiscriminate use of the atmosphere as a gigantic sewer, however, has led to isolated severe air pollution episodes and ultimately to a global deterioration of the quality of the ambient air. Technological advancement, industrial expansion, population explosion, urbanization, and man's constant striving for a higher standard of living—as measured in terms of the number and value of his material possessions—have all contributed to the creation of the current air pollution crisis.

Technology and industry are integral parts of an economy which follows the philosophy that natural resources should be exploited as fast as possible, quickly converted into not-especially-durable goods, then rapidly distributed to individual consumers who have been convinced that they need these goods but who have not been told just what to do about the wastes that these goods inevitably produce.

The automobile industry presents the epitome of this vicious cycle. In order to transport his 160 pounds over the 20 miles to his working place, man, the consumer, is offered his choice from a multitude of 4,000-pound vehicles propelled by 300 horsepower engines. Not only do these luxurious monstrosities waste his natural resources and necessitate the covering of significant portions of his world with asphalt and concrete, they also annually emit into the air over the United States alone more than 60 million tons of carbon monoxide, 10 million tons of hydrocarbons, 3 million tons of nitrogen oxides, and substantial quantities of other poisons such as lead and asbestos. This same economic philosophy has driven man to migrate from the rural farmlands to the urban complexes in search of his "higher standard of living." Through overpopulation, poor city planning, and his inability to deal with his own pollution, man has managed to create unbearable living conditions right in the middle of his most desirable land areas. This situation has spawned the sprawling suburbs. In attempting to alleviate his congestion problems, however, man has inadvertently aggravated his air pollution problems. Now separate pollution trails converge from all directions to make the air over cities not only unpleasant but deadly.

Yet one hears man say to himself, The price for progress and urbanization is polluted air, or It is necessary to put up with a certain amount of polluted air in order to maintain the present standard of living. These statements indicate that man's mind has been polluted at least as much as the air he breathes.

But man's environmental awareness has been increasing within recent years. He is slowly beginning to realize that a truly high standard of living can be achieved only if he can find some way to improve the quality of both his own life and his surroundings as well.

Man can no longer afford to support his own apathy. He must begin now to restructure his economic and social philosophies and to redirect his technological efforts towards the conservation and preservation of one of the most rapidly disappearing natural resources—breathable air.

CHAPTER 1

DEFINITION, SOURCES, AND TYPES OF AIR POLLUTION

The statement is basically true that the atmosphere has always been polluted to some degree. During certain geologic epochs natural pollution from volcanic eruptions has had global implications in that it changed the earth's climate. On a smaller, but still area-wide scale, natural pollution from grass fires and odors from foul-smelling swamps and marshes must have caused at least local nuisances.

Man-made pollution has existed as a local problem since man invented fire for cooking and for warming his ill-ventilated caves. Historians report crude oil combustion in Persian shrines as early as 500 B.C. In his poems Horace deplored the smoke-blackened temples of Rome in 100 B.C. In medieval times man-made pollution became such a menace that British kings decreed that fouling of the London air by smoke was an offense which was punishable by the death penalty through hanging.

DEFINITION OF AIR POLLUTION

Man-made air pollution produced at relatively small urban and industrial agglomerations has within the past century developed into a real threat to life. Rossano [1] distinguishes three major categories of man-made pollution:

1. *Personal.* This comprises all gases and particles from cigarette, cigar, and pipe smoking, and the use of household sprays. Smoking has been shown to be hazardous to health, and sprays, carelessly applied as advertised, can be extremely detrimental. The solution to this pollution problem is seemingly quite simple: do not smoke and

do not permit others to smoke in your presence; do not use sprays indiscriminately.

2. *Occupational.* This category comprises all exposure of persons to pollution at their working place. Industrial hygienists have studied occupational diseases for more than a century and have developed MAC values (maximum allowable concentrations). These threshold limits can not, however, be applied to the third category, community air pollution, because they relate to healthy young male workers and to a limited exposure time.

3. *Community.* Ambient or outdoor air pollution is usually considered as community air pollution which nobody can escape. Largely unpredictable human and meteorologic factors set the levels of community pollution. When people think of pollution, they usually have in mind community air pollution, which can strike in the form of short-term disasters causing thousands of deaths, or as long-term influences changing the climate of the planet. It is, however, the chronic effects from all three categories that have to be studied more carefully if preventive measures are to be successfully applied. The legislation, although specific, is also somewhat one-sidedly oriented towards community pollution when it defines air pollution as "The presence in the outdoor atmosphere of any form of contaminant . . . inimical or which may be inimical to the public health, safety, or welfare or which is, or may be injurious to human, plant or animal life, or to property, or which unreasonably interferes with the comfortable enjoyment of life or property" [2].

SOURCES OF AIR POLLUTION

Man-made air pollution sources can be conveniently grouped under single or point sources, multiple or area sources, and line sources. Point sources such as steel mills, power plants, oil refineries, and pulp and paper mills, etc., with their tall stacks are usually identified as major contributors to air pollution. Equally bad because of sheer quantity and low levels of emissions are the area sources; residential areas, apartments, office buildings, hospitals, and schools are the greatest contributors. Line sources, such as expressways, so far seem to affect only the drivers. However, in the narrow, canyon-like streets of the cities the automobile constitutes a great health hazard to the general public, because it emits not only the largest quantities of pollutants of any type of source, but also because it emits the poisons at breathing level. The U.S. Public Health Service has cataloged the five major sources of pollutants (Table 1.1). The values given in millions of tons refer to the United States of 1965.

Motor vehicles. It is quite obvious that by far the largest amount of pollutants (60 percent) were emitted by the 90 million motor

Table 1.1 Major sources of air pollutant emissions in the United States in millions of tons per year, 1965

Source	Carbon monoxide	Sulfur oxides	Hydro-carbons	Nitrogen oxides	Particulate matter	Total	Percent of total
Motor vehicles	66	1	12	6	1	86	60
Industry	2	9	4	2	6	23	17
Power plants	1	12	1	3	3	20	14
Space heating	2	3	1	1	1	8	6
Refuse disposal	1	1	1	1	1	5	3
Total	72	26	19	13	12	142	100

Source: USDHEW, PHS, Washington, D.C., 1966 [3].

vehicles registered in 1965. In 1970 the 100 million mark was passed, which means of course more polluted highways and roads and more congestion. Hydrocarbons and nitrogen dioxide emanating mainly from motor vehicles are the major ingredients of photochemical smog. It is now a fact that downtown areas of all our cities have a smog problem. Automobiles also emit sizable quantities of such poisons as lead, arsenic, and aldehydes.

Industry. About one-fifth of the total air pollution in the United States is produced by such industrial contributors as metallurgical plants and smelters, chemical plants and petroleum refineries, fertilizer and synthetic rubber manufacturers, pulp and paper mills, etc. Industry has always been proud of its enormous contribution to technological progress and to raising the standard of living although a certain price had to be paid. This success philosophy must now be evaluated in the context of the overall public welfare. It is recognized that standard of living cannot be equated with quality of life. Industry has a right to produce and make profit, but it has no right to ruin the public's environment.

Power plants. In the United States still more than 90 percent of the electrical energy is generated by burning coal and oil. Because sulfur is one of the major ingredients of these fuels, power plants are the greatest contributors to sulfur dioxide (SO_2) pollution (see Table 1.1). The electric power consumed per person per year in kilowatt-hours in 1950 was 2,000; and in 1968 it was 6,500. It is estimated to rise to 11,500 in 1980, and to about 25,000 by the year 2000 [4]. Power consumption increases twice as fast as the population. The use of low-sulfur fuel, which is rather scarce, is one of the methods used to reduce sulfur emissions. Other methods of power production, such as hydroelectric power, energy from gas turbines, magnetohydrodynamics, garbage burning, and nuclear reactors, should be investigated for their economic feasibility and potential hazards.

Space heating. Emissions from area sources constitute the fourth largest source of pollution (see Table 1.1). These include single and multiple houses, apartment houses, offices, stores, restaurants, hotels, clubs, hospitals, schools, laundries, and dry cleaners. Residential areas built far away from industrial pollutors and separated from heavy traffic are usually their own worst pollutors with open fireplaces in the cold months and barbecues in the warm summer months.

Refuse disposal. The annual production of waste material in the United States is about 150 million tons; most is burned in municipal, industrial, and commercial incinerators, in apartment house incinerators, or in open dumps. An ordinary apartment house incinerator emits about 25 pounds of pollutants for every ton of refuse burned [5]. Some of it is buried pollution-free in sanitary landfill areas. Trash, leaves, and tire burning in a residential area is one of the most vicious ways of spoiling a whole neighborhood.

The list of sources given above would not be complete without mentioning the emissions from railroad engines, ships, and aircraft.

Transportation services. The general policy with railroad diesel engines is not to shut them off, but to put them on a side track when not in use. At night, when all dispersion is subdued, a local buildup of high SO_2 concentrations can occur. Ships are considerable contributors of smoke and SO_2 to the pollution level of port cities. As the cities grow, their airports also undergo expansion. Increased traffic volume and larger planes produce more traffic holdups and in-line queuing on the ground, and cruising within the airports' waiting corridors. This is a waste of fuel energy, the passenger's money, and the public's air. Table 1.2 compares pollutant emissions

Table 1.2 Comparison of average daily emissions from the combustion of fuels by motor vehicles, power plants, and jet aircraft in Los Angeles county, 1969

Pollutants	Motor vehicles	Power plants Apr. 15-Nov. 15	Power plants Nov. 16-Apr. 14	Jet aircraft
Particulates	43	1	6	11
Carbon monoxide	9,282	Neg.	Neg.	24
Nitrogen dioxide	624	135	145	7
Hydrocarbons	1,677	4	6	61
Sulfur dioxide	31	30	115	3
Totals	11,657	170	272	106

AVERAGE DAILY EMISSIONS (TONS PER DAY)

Source: R. E. George, et al., 1969 [6].

from automobiles, power plants, and jet aircraft for Los Angeles County. Automobiles are naturally the biggest pollutors, but emissions from jet aircraft are coming very close in quantity to those from power plants.

Commercial and agricultural activities. Demolition, construction, and spray painting are the most common sources of pollution within this category. Field burning and dusting and spraying of the fields with pesticides are some of the methods farmers use to contribute their share to the general pollution problem.

Minor sources of air pollution. If one compares the total emissions of 142 million tons of pollutants from major pollution sources in Table 1.1 with the 102 million tons of emissions from so-called minor pollution sources in Table 1.3 [7], then the term minor pollution sources is perhaps a misnomer. The amounts of organic

Table 1.3 Minor sources of air pollutant emissions in the United States in thousands of tons per year, 1965

Source	Total emissions	Percent of total
Organic compounds from vegetation	70,200.0	68.6
Ground dust	30,000.0	29.3
Salt spray from oceans	1,226.4*	1.2
Spray aerosols from cans	380.0	0.4
Rubber from vehicle tires	301.4	0.3
Cigarettes	116.8	0.1
Hydrogen sulfide from natural sources	95.3	< 0.1
Organic compounds from perfumes	35.0	< 0.1
Cosmic dust	23.3	< 0.1
Total	102,378.2	100.0

*Total emission of salt spray distributed within 200 miles of coast.
Source: Calculated from data given by V. J. Marchesani, et al., 1970 [7].

compounds from vegetation, such as pollen and other aero-allergens which produce, for example, the blue hazes of the Blue Ridge and Smoky Mountains, compare in quantity with those emitted from motor vehicles. The decisive difference is that these natural pollutants are spread over large and mostly uninhabited areas, whereas most automobile pollution is emitted in the most densely populated areas. The large tonnage of gaseous and particulate aerosols from aerosol sprays and smoking could also be of concern, because it is emitted into rather small and confined spaces.

Natural sources of air pollution. The natural pollution sources shown in Table 1.3 have to be supplemented. During catastrophes large amounts of gases and ash from volcanic eruptions and gases and

8 ATMOSPHERIC POLLUTION

smoke from forest, grass, and swamp fires blacken the skies and increase the background pollution levels for years, even at points far distant from the original source. Pollutants and malodors from microorganisms, gases and odors from swamps and marshes, natural radioactivity, fog and mist, and ozone from the ozonosphere and lightning are more constant causes of natural air pollution.

MAJOR TYPES AND CONCENTRATIONS OF AIR POLLUTION

There are hundreds of different air pollutants. Only the major types, as presented in Tables 1.1 to 1.3, can be discussed in detail. Besides giving the total tonnage of pollutant emissions, it is often important to know the amounts of emissions in proportion to the amount of air, i.e., the number of parts of a pollutant for each million or billion parts of air (ppm or ppb), or the number of micrograms (1 millionth of a gram) of a pollutant per cubic meter of air ($\mu g/m^3$). This information is important when danger levels for people are considered, when levels of adverse effects for animals, vegetation, and property are studied, and when control standards are proposed.

A good idea of how much man-made pollution we have already accumulated in the atmosphere can be obtained by comparing the composition of clean air (Table 1.4) with the composition of polluted city air (Table 1.5). The single most common element is, of

Table 1.4 Composition of clean, dry air near sea level

Component	Concentration (ppm)	Component	Concentration (ppm)
Nitrogen	780,900	Methane	1.5
Oxygen	209,400	Hydrogen	0.5
Argon	9,300	Carbon monoxide	0.1
Carbon dioxide	318	Ozone	0.02
Neon	18	Nitrogen dioxide	0.001
Helium	5.2	Sulfur dioxide	0.0002

Source: Report, American Chemical Society, 1969 [8].

course, nitrogen with 78 percent (parts per hundred), which amounts to 780,900 ppm (parts per million). The most common pollutants, such as carbon monoxide (CO), ozone (O_3), nitrogen dioxide (NO_2), and sulfur dioxide (SO_2), are present in minute concentrations in a natural atmosphere. This picture of equilibrium can be easily upset in a city atmosphere, when for instance the two-year mean SO_2-concentration for Chicago can be 675, and that for a five-minute maximum value can be 8,200 times higher than the background SO_2-concentration (see Table 1.5). In any discussion on pollution concentrations it is obviously of importance whether one looks at mean annual trend values, at seasonal variations, at weekly or diurnal changes, or at short-period peak values (Table 1.5). Total tonnage values of pollutants may be impressive, but it is the short-period concentrations that are of greatest significance in health and air pollution control. Methane and ozone in Table 1.4 are only somewhat comparable to total hydrocarbons and total oxidants in Table 1.5. Finally, absolute magnitudes of concentrations of different pollutants should not be compared, because, for example, the threshold value of CO is 50 ppm and that of SO_2 is much lower, namely 5 ppm. We shall now look more closely at the most common individual pollutants.

Table 1.5 Composition of polluted city atmospheres, 1962-1963 CAMP data

City	Days of valid data	Two-year mean	Max. month	Max. day	Max. hour	Max. 5-min.
Sulfur dioxide						
Chicago	618	0.135	0.344	0.79	1.36	1.64
Cincinnati	605	0.029	0.056	0.11	0.48	0.99
New Orleans	472	0.010	0.020	0.06	0.15	0.31
Nitrogen dioxide						
Chicago	646	0.042	0.064	0.13	0.22	0.26
Cincinnati	623	0.030	0.043	0.09	0.25	0.30
New Orleans	525	0.019	0.031	0.05	0.13	0.15
Total oxidant						
Chicago	459	0.004	0.010	0.07	0.22	0.25
Cincinnati	555	0.014	0.040	0.09	0.20	0.24
New Orleans	473	0.019	0.038	0.08	0.18	0.19
Total hydrocarbon						
Chicago	562	3.2	4.3	6	12	19
Cincinnati	517	3.3	4.6	10	17	25
New Orleans	243	1.8	3.1	5	14	18
Carbon monoxide						
Chicago	244	7.6	9.6	19	36	50
Cincinnati	146	6.9	10.4	16	22	31
New Orleans	53	4.4	4.8	13	36	43

Source: D. A. Lynn and T. B. McMullen, 1966 [9].

Oxides of sulfur. Sulfur (S) is an impurity in coal and in fuel oil. Through combustion it enters the atmosphere as sulfur dioxide (SO_2), hydrogen sulfide (H_2S), sulfurous (H_2SO_3) and sulfuric acid (H_2SO_4), and various sulfates. A city the size of Cincinnati emits into the atmosphere annually more than 7,000 tons of sulfur oxides

(SO_x), New York City about 1.5 million tons of SO_2, Great Britain about 5.8 million tons of SO_2, the United States 26 million tons of SO_2 [3] and the whole world about 80 million tons of SO_x, which is more than twice the world S production of about 30 million tons. A continued waste of such an important raw material is irresponsible. The 80 million tons of SO_2 per year equally distributed over the globe would increase the world SO_2-concentration by about 0.006 ppm. Fortunately precipitation removes all acid and sulfates within a period of about 43 days [10].

Carbon monoxide (CO) and carbon dioxide (CO_2). Carbon monoxide, a colorless, odorless, and lethal gas, results from incomplete combustion of carbonaceous materials. Of the world's total CO production of 232 million tons, 80 percent is produced by automobiles. If this amount were evenly spread over the lower atmosphere, it would increase the CO content by 0.03 ppm per year. This is very significant, because CO is a very stable gas. In an experiment a CO-O mixture under exposure to sunlight remained unchanged even after seven years. There are four possible ways to get rid of this deadly gas: it may eventually escape into the general atmosphere; it may oxidize to CO_2; it may be used by metabolizing bacteria; or it may be absorbed by the oceans.

Carbon dioxide, a heavy, colorless, and odorless gas, is generally not considered an air pollutant, because it is essential in all life processes. In industrial areas CO_2 concentrations have already reached the 1,000 ppm level as compared with the normal level of 318 ppm for clean air (see Table 1.4). Increased CO_2 levels produce the so-called greenhouse effect which raises the temperatures in cities. Since the industrial revolution some 330 billion tons of CO_2 from combustion processes have been added to the atmosphere. This amounts to about 14 percent of the natural CO_2 content in the air. Since CO_2 released into the air remains there for at least several centuries [11], we have reason to be greatly concerned. The impact CO_2 has on changing the global climate is discussed in Chapter 2.

Hydrocarbons (HC). Hydrocarbons originate from the combustion of gasoline, coal, oil, natural gas, and wood, from evaporation of gasoline and industrial solvents, and from natural sources, mainly the decomposition of vegetation. Man-made HC emissions amount to 90 million tons a year, whereas flooded swamps produce HC at a rate of 1.6 billion tons a year. Again, however, it must be remembered that man-made HC are emitted into the rather small air space in which most of mankind is crowded together.

In air pollution control the unsaturated HC of the olefin group and the compounds belonging to the aromatic or benzene group are of greatest concern. The unsaturated olefins react easily with other

DEFINITION, SOURCES, AND TYPES OF AIR POLLUTION 11

chemicals. For instance, only a few tenths of 1 ppm of nitrogen oxides (NO_x) and less than 1 ppm of unsaturated and hence highly reactive HC together with sunlight have been found to initiate the Los Angeles type smog. Favorable conditions for smog formation are given in most of our cities (compare the concentrations of HC and NO_x in Table 1.5). The aromatics have been found to be carcinogenic or cancer-producing. The most powerful of these is benzpyrene (also written 3,4 benzpyrene or benzo(a)pyrene). Ozone and NO_2 are the principal removal processes for HC. The possible residence time of HC in the air is still largely unknown. A time as long as 20 years has been suggested.

Oxides of nitrogen (NO_x). Nitric oxide (NO), a relatively harmless gas, turns into a pungent, yellow-brown, harmful gas when oxidized to nitrogen dioxide (NO_2). Man-made NO_2 originates from stationary sources, such as fertilizer and explosives industries, and from mobile sources, such as automobiles, trucks, and buses. As described under hydrocarbons, NO_2 in small concentrations is an essential ingredient in photochemical smog formation. The major natural sources for NO_x compounds are organic decomposition in the soil and perhaps in the ocean. The amount of NO_x which does not react photochemically is usually removed from the air within about three days [8].

Photochemical products. In the presence of reactive HC (olefins) solar energy is absorbed by NO_2 to form photochemical smog. During this process nitric oxide and atomic oxygen are formed. The atomic oxygen (O) reacts with the usually present oxygen molecules (O_2) to form the colorless and pungent gas ozone (O_3). Since O_3 is an early and continuing product in the smog formation, and since the presence of O_3 keeps the oxydizing process going, O_3 is almost interchangeably used with the term oxidant, which is a measure of how much smog formation is taking place. Hundreds of chemical processes occur as long as there is sufficient supply of HC, NO, and NO_2 from automobiles, and O_3 and solar radiation. Some of the better known irritating photochemicals are PAN (peroxyacetyl nitrate) and aldehydes. A simplified reaction scheme for photochemical smog can be found in a report of the American Chemical Society, p. 38, [8], and a complete smog formation theory is presented by Haagen-Smit [12].

Particulate matter. It consists of solid and liquid particles of a wide range of sizes varying from greater than 100 μm to less than 0.1 μm (1 μm or micrometer = 10^{-4} or 1/10,000 cm). Particles larger than 10 μm consist mainly of dust, coarse dirt, and fly ash from industrial and erosive processes. These large particles usually settle out rapidly.

They are collected in dustfall jars, and the values given in tons per square mile per month range from 22 for Los Angeles to 72 for Detroit [13]. Particles below 10 µm remain much longer as suspended matter in the air. Particulates below 5 µm are known as smoke and fume, those under 1 µm as aerosols. Suspended particulates are either measured by high-volume samplers or spot tape samplers.

Table 1.6 Relative severity of pollution in 65 urban and industrial areas of the United States, 1967

1. New York	34. Nashville
2. Chicago	35. San Francisco-Oakland
3. Philadelphia	36. Seattle
4. Los Angeles-Long Beach	37. Lawrence-Haverhill
5. Cleveland	38. New Haven
6. Pittsburgh	39. York
7. Boston	40. Springfield-Chicopee-Holyoke
8. Newark	41. Allentown-Bethlehem-Easton
9. Detroit	42. Worcester
10. St. Louis	43. Houston
11. Gary-Hammond-East Chicago	44. Chattanooga
12. Akron	45. Memphis
13. Baltimore	46. Columbus, Ohio
14. Indianapolis	47. Richmond
15. Wilmington, Delaware	48. San Jose
16. Louisville	49. Portland, Oregon
17. Jersey City	50. Syracuse
18. Washington, D.C.	51. Atlanta
19. Cincinnati	52. Grand Rapids
20. Milwaukee	53. Rochester
21. Paterson-Clifton-Passaic	54. Reading
22. Canton	55. Albany-Schenectady-Troy
23. Youngstown-Warren	56. Lancaster
24. Toledo	57. Dallas
25. Kansas City	58. Flint
26. Dayton	59. New Orleans
27. Denver	60. Fort Worth
28. Bridgeport	61. San Diego
29. Providence-Pawtucket	62. Utica-Rome
30. Buffalo	63. Miami
31. Birmingham	64. Wichita
32. Minneapolis-St. Paul	65. High Point-Greensboro
33. Hartford	

Source: *Science*, Aug., 1967 [15].

Depending on the size of the city mean annual concentrations of a few hundred µg/m³ and instantaneous concentrations of several thousand µg/m³ in pollution episodes are quite common.

The life of particles in the troposphere lasts only a few days [14]. If they are, however, injected into the stratosphere, they may hover around the globe for several years. This may have a severe impact on the global climate.

After the discussion of the major sources and types of air pollution it would be quite interesting to know where the pollution situation is worst. The National Center for Air Pollution Control, U.S. Public Health Service, has given the relative severity of air pollution for 65 urban and industrial areas (Table 1.6). The tabulation is based on concentrations of suspended particulates and SO_2, on gasoline consumption, and emissions from automobiles [15].

SUMMARY

A short historical review shows that man has lived in an atmosphere which has always been polluted to some degree. To natural pollution from volcanic activity, grass and forest fires, man added his portion from cooking and heating. It was not, however, until the onset of the industrial revolution and large-scale urbanization that air pollution developed into a regional and continent-wide problem.

There are three general types of man-made pollution, namely personal, occupational, and community air pollution. The emphasis is on community or ambient man-made pollution which is produced by single or point sources, multiple or area sources, and by line sources. The major contributions to these sources stem from motor vehicles and other transportation means, industry, power plants, space heating, refuse disposal, and commercial and agricultural activities.

The composition and concentrations of substances in a clean atmosphere are compared with those of polluted city air. The major types of pollutants, such as the oxides of sulfur and nitrogen, carbon monoxide and dioxide, hydrocarbons, photochemical oxidants, and particulate matter, are discussed. Estimated residence time in the atmosphere and typical ground concentrations of these pollutants are given.

REFERENCES CITED

[1] A. T. Rossano, Jr., "Sources of Air Pollution" in A. T. Rossano, Jr. ed., *Air Pollution Control Guidebook for Management*, Env. Scienc. Serv. Div., Stamford, Conn., 1969.
[2] Air Pollution Control Act, Jan. 8, 1960, Section 3(5), as amended, 35 P.S. 4003(5).
[3] USDHEW, *The Sources of Air Pollution and Their Control*, PHS Publ. No. 1548, Washington, D.C., 1966.
[4] D. E. Abrahamson, *Environmental Cost of Electric Power*, A Scientists' Institute for Publ. Inf. Workbook, New York, 1970.
[5] E. Edelson, *The Battle for Clean Air*, Publ. Affairs Pamphlet No. 403, New York, Aug., 1967.
[6] R. E. George, et al., "Jet Aircraft: A Growing Pollution Source," *JAPCA* 19(11), 847-855, Nov., 1969.
[7] V. J. Marchesani, et al., "Minor Sources of Air Pollutant Emissions," *JAPCA* 20(1), 19-22, Jan., 1970.
[8] *Cleaning our Environment*, Rept. Am. Chemical Soc., Washington, D.C., 1969, p. 23 ff.
[9] D. A. Lynn and T. B. McMullen, "Air Pollution in Six Major U.S. Cities as Measured by the Continuous Air Monitoring Program," *JAPCA* 16(4), 186-190, Apr., 1966.
[10] C. E. Junge and R. T. Werby, "The Concentration of Chloride, Sodium, Potassium, Calcium, and Sulfate in Rain Water over the U.S.," *J. of Meteor.* 15, 417-425, 1958.
[11] C. D. Keeling, "The Concentration and Isotopic Abundances of CO_2 in the Atmosphere," *Tellus* 12, 200-203, 1960.
[12] A. J. Haagen-Smit, "The Chemistry and Physiology of Los Angeles Smog," *Industr. Engineering Chem.* 44, 1342-1346, 1952.
[13] H. Landsberg, *Physical Climatology*, 2nd ed., Gray Printing Co., Inc., DuBois, Pennsylvania, 1962, p. 318.
[14] C. E. Junge, *Air Chemistry and Radioactivity*, Academic Press, Inc., New York, 1963, 382 pp.
[15] "Air Pollution: Where the Problems Are Worst," *Science* 157, 785, Aug., 1967.

CHAPTER 2

METEOROLOGICAL ASPECTS OF AIR POLLUTION

The Influence of Weather on Air Pollution

Although in most cases we have the technological ability to control air pollution, we still lack the legal authority to carry it through. Therefore, viewed realistically, our atmosphere will be used as a giant sewer for some time to come. The layer into which the contaminants are emitted, on the average only about 12 kilometers deep, seems to be at first glance a gigantic air reservoir of about 5×10^{18} m³. However, most of the pollution is emitted into relatively small airsheds over urban and industrial areas. Episodic air pollution disasters have shown that at certain periods of stagnating air masses the limit of the dispersive and cleansing power of the atmosphere has already been reached. Logically therefore, a self-preserving exploitation of this natural resource air can only be made if one understands the meteorological processes of dispersion and removal of pollutants and incorporates them into an air resources management and control system.

All decisive weather processes which influence pollution dispersal take place in a very thin air layer, the troposphere, the depth of which is to the earth as is the skin of a peach to the peach itself. Of this shallow air layer, into which millions of tons of pollutants are emitted each year, only the lowest few thousand meters are suitable at all to sustain life. If it were not for the natural dispersing, diluting, scavenging, and removal processes in the troposphere, the air on earth would perhaps no longer be breathable. The energy for all these life-sustaining meteorological processes is supplied by the sun.

SOLAR ENERGY

Energy is transmitted from the sun to the earth in the form of electromagnetic waves called solar radiation. Depending on the pollution loading of the atmosphere, a good deal of the solar beam does not reach the earth's surface, but is rather attenuated or scattered back to space (Figure 2.1). On its 150 million kilometer journey, the solar beam loses hardly any energy (only about 20 percent) until it reaches the lowest few thousand meters of city air. Thus, for instance, on a polluted day, the air over Cincinnati below about 3,500 feet can reduce the incoming solar energy by more than 60 percent.

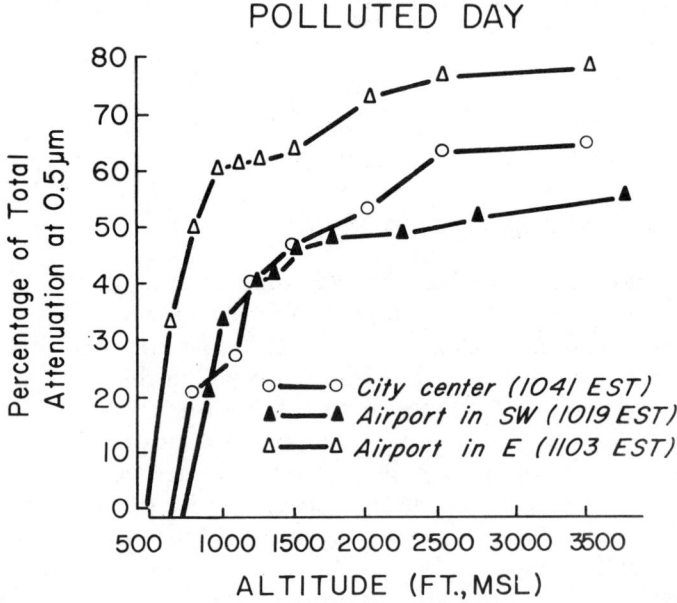

Figure 2.1 Cumulative frequency graphs of solar attenuation at 0.5 μm as a function of altitude, 11/6/69.

Despite this enormous reduction of solar radiation the energy seems to be sufficient to initiate a photochemical process resulting in the Los Angeles type smog. The complicated processes involved in the photochemical reactions between nitrogen oxides, hydrocarbons, and solar energy are described in Chapter 1. When photochemical damage to plants was first discovered in Los Angeles County in the 1940s it only covered a few square kilometers; today more than 30,000 square kilometers of California land report smog damage [1]. Smog is now found in all cities of the United States.

METEOROLOGICAL ASPECTS OF AIR POLLUTION 17

The remaining part of the shortwave solar radiation, which is not scattered, is absorbed by the earth's surface and reemitted as longwave or heat radiation. The air layers closest to the radiating surface are heated and start to rise as convectional air streams. The air motion produced in this manner will disperse and dilute air pollutants. It is ultimately the energy from the sun that determines the two major meteorological processes responsible for high or low pollution levels, namely atmospheric stability and air flow.

STABILITY CONDITIONS

Solar radiation by heating the ground surface and the air adjacent to it produces changes in temperature with height, termed lapse rate. In well-mixed dry air, the dry adiabatic lapse rate is 1° C per 100 meters or 5.4° F per 1,000 feet, which means that for each 100 meter increase in altitude the air temperature decreases by 1° C (Figure 2.2). When the temperature decreases with height, as it normally does, the lapse rate is said to be positive, and a good dispersion and vertical transport of pollutants can be expected under unstable weather conditions. The air is said to be stable when the temperature increases with height, i.e., when the lapse rate is negative or an inversion. Such conditions prevent any vertical dispersion and trap the pollutants within the inversion layer.

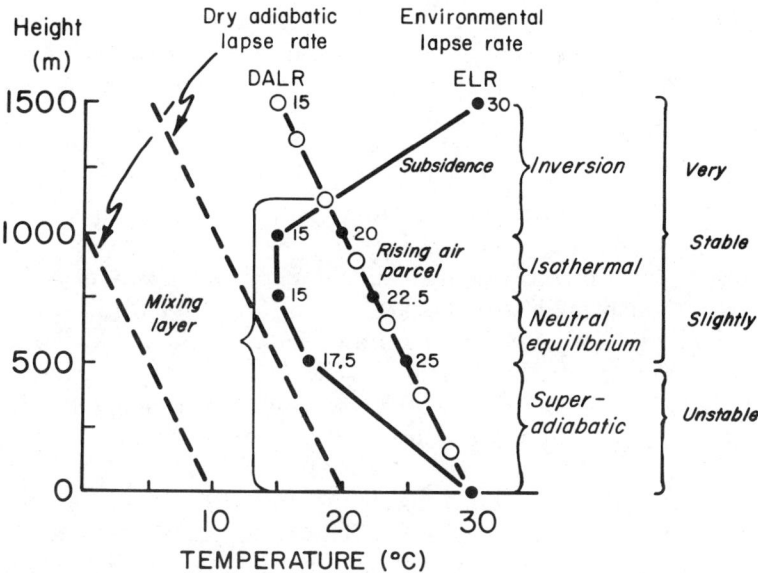

Figure 2.2 Stability and instability in dry air.

There are three major processes that cause air to rise and the pollutants accumulated in it to disperse: 1. convective air flow, which is produced by heating of the earth's surface through solar radiation; 2. orographic ascent, in which air masses are forced to rise over hills and mountains; 3. air mass lifting, when, for example, warm air slides over cold air along a warm front. The example of convective air flow has been chosen to demonstrate the intricate relationship between stability, vertical air flow, and pollutant dispersion (see Figure 2.2). The data for the environmental lapse rate (ELR) plotted on the temperature-height diagram are usually obtained from radiosondes, balloon-borne instruments, transmitting temperature, humidity, and pressure readings to a ground station. The ELR represents the actual temperature of the ambient air at the indicated heights. The dry adiabatic lapse rate (DALR) of 1° C per 100 meters is a theoretical rate at which the temperature of an air parcel cools when forced to rise. At the surface DALR and ELR are the same. Strong solar heating makes the air layers close to the surface rise and cool off. At 500 meters the rising air is 25° C and thus 7.5° C warmer than the ambient air (17.5° C). The air mass is very unstable and under superadiabatic conditions has a great tendency to rise and disperse pollutants.

Within the layer of neutral equilibrium (500-750 meters) the air would normally have neither a tendency to rise nor to fall and thus is rightly termed slightly stable. But in this example, due to the initial strong heating, the rising air is still 7.5° C warmer than the ambient air, so that its inherent buoyancy will let it continue to rise.

Within the isothermal layer (750-1,000 meters), which is actually quite stable, the rising air at 1,000 meters is still 5° C warmer than its surrounding air so that it will continue to rise. At 1,125 meters the temperature of the rising and the ambient air is the same. The rising air parcel in very stable surroundings will now start to sink back. Within the 1,000 to 1,500 meter air layer we find a sinking or subsiding air motion producing a subsidence inversion. Our air parcel theoretically displaced to a height of 1,500 meters would be 15° C colder than the ambient air and immediately sink back. Thus, the subsiding air masses will effectively seal off any further penetration of the rising air so that in this example vertical transportation and dispersion of pollutants is possible only within a mixing layer 1,125 meters deep.

The example could most likely be found on a sunny summer afternoon in a temperate climate. The synoptic (large-scale simultaneous) weather situation would be that of an anticyclonic (high pressure) weather system with its characteristic subsidence inversions under sunny skies.

Another type of stability condition is the surface-based shallow radiation inversion, which may be produced nightly under clear skies and light winds through outgoing longwave radiation. High pollution concentrations accumulate beneath the inversion lid. The following morning, when solar heating breaks up the inversion, large amounts of pollutants are brought down to breathing level in a process termed fumigation (see Figure 2.4, No. 6). If the inversion layer is deep, however, and the usually formed fog is dense enough to effectively prevent solar heating of the ground, then the polluted inversion layer may persist continuously for several days. Only a change in weather, such as the passage of a frontal system, can bring relief. All major air pollution disasters discussed in Chapter 3 have occurred under stable stagnating air masses with shallow inversions.

AIR FLOW AND TURBULENCE

Differential solar heating of the earth's surface produces pressure and temperature differences and hence air motion. Basically, wind transports air pollutants from one place to another, whereas turbulence dilutes them [2]. Wind speed and turbulence are proportional to the transport and dispersion of pollutants in the sense that the higher the wind speed and turbulence the lower will be the concentration of pollutants. Wind direction only influences the direction of transport and the area of spread of the pollutants.

The processes of transportation, diffusion, and dilution through wind and turbulence are clarified in Figure 2.3. If the size of an eddy (a constituent of the random turbulent motion) is larger than the size of the plume, then the plume will be transported downwind by the eddy. If, on the other hand, the eddy is smaller than the plume, the eddy will diffuse it. The process of dilution is further exemplified in Figure 2.3. Suppose a stack is emitting a certain pollutant at a rate of 20 grams per second, and the wind speed is 1 meter per second. Since 1 meter of air passes by the stack each second, a downwind plume 1 meter in length will contain 20 grams of that pollutant. At the same emission rate but at a wind speed of 5 meters per second each meter of plume length would then only contain 4 grams of pollutants.

There are two types of turbulence or eddy diffusion. Mechanical turbulence is produced by air passing over a rough surface, and thermal turbulence is induced by thermal heating and convectional (vertical) air flow. Both types of turbulence help to disperse pollutants.

The most important difference between vertical and horizontal air motion and hence pollutant dispersion is that of magnitude. In the vertical, dispersion is restricted to about 12 kilometers. But in the horizontal, the entire globe is actually available. The global wind pattern of prevailing westerlies, or northeast polar and trade winds etc., is locally modified by such mesoscale winds as valley and slope or sea and land breezes. The country breeze is initiated by the heat absorbing cities (heat-island effect). All these local winds can bring relief by replacing some of the polluted air masses that have become stationary under high pressure systems with little or no air motion.

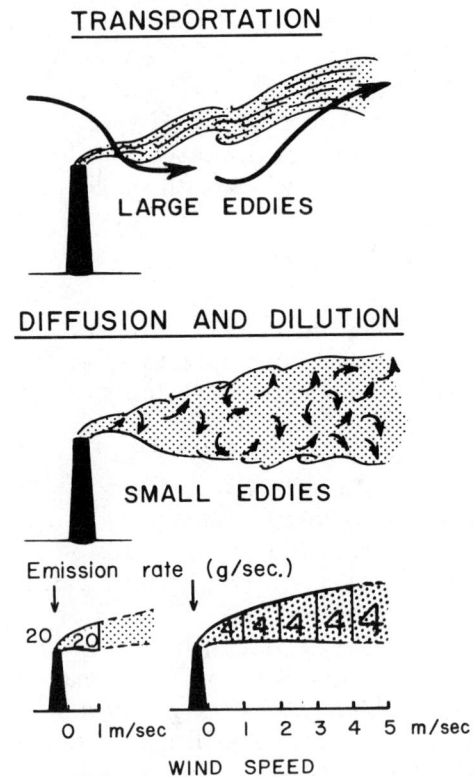

Figure 2.3 Transportation, diffusion, and dilution of air pollution.

SCAVENGING PROCESSES

In Chapter 1 some estimations of residence times of the major pollutants in the atmosphere were presented. But we still do not know at which rate pollutants are removed from the atmosphere. We also do not know whether the present rate of pollution emission on a global basis exceeds the rate at which they are being removed from the atmosphere [3]. One fact is clear: without the cleansing processes the earth would be uninhabitable.

Particulates of a size greater than 10 μm are quickly removed from the atmosphere by gravitational settling. Particles in the size range of 0.1 to 10 μm act as condensation nuclei. By coagulation and attachment of water vapor the particles grow in size and then settle out. But precipitation in the form of rain, drizzle, or snow is the most effective cleanser of the atmosphere. Particles are either intercepted by falling raindrops (washout) or by raindrop formation within clouds and subsequent falling as rainfall (rainout) [4]. A uniform rainfall of 1 millimeter per hour over a 15-minute period can scavenge 28 percent of the 10 μm particulates from a volume of air through which the rain passed [5]. For particles of 2 μm and smaller the scavenging through precipitation becomes insignificant. Photochemical reactions and oxidation also help to remove pollutants by chemical conversion. Meteorological factors such as solar radiation and humidity are essential in either removal processes.

APPLICATION OF METEOROLOGY TO AIR POLLUTION

The geometrical forms of stack plumes are a function of the vertical temperature and wind profiles. Experienced observers can predict the plume type by looking at the temperature and wind profiles; vice versa, by looking at a plume, they can estimate the stability conditions and the dispersive capacity of the atmosphere. Careful analysis of Figure 2.4 and practice can make an expert of you.

Plume behavior. A very common type of plume behavior is *looping*. It occurs under superadiabatic conditions with moderate to high wind speeds and large convective (vertical) mixing. For low stacks, looping can produce high surface concentrations. This plume type can occur under anticyclonic (high pressure) and cyclonic (low pressure) weather types.

Coning can occur both during day and night and in all seasons. Under cloudy and windy weather with prevailing mechanical turbulence, either slightly unstable or slightly stable conditions produce an evenly shaped plume. This plume type fits almost ideally the Gaussian diffusion model discussed in the following section, and does not contribute significantly to pollution near the ground.

Fanning occurs most often at night and in the early morning in all seasons. Anticyclonic weather with cloudless skies, light winds, and heavy radiation loss from the surface produces a very stable radiation inversion layer near the ground. Pollutants emitted during such atmospheric conditions fan in a flat, straight ribbon downwind, or meander if the wind direction changes. For high stacks, fanning is considered a favorable meteorological condition for stack releases, because the effluent stays aloft and does not contribute to ground pollution. If, however, the stack is short relative to the surrounding topography, then a serious pollution problem may result.

22 ATMOSPHERIC POLLUTION

Lofting may occur in the early evening, at night, and in the early morning. Under anticyclonic weather with cloudless calm nights a shallow radiation inversion forms near the ground below the effluent emission height. Slightly unstable conditions aloft disperse the pollutants upwards. This is the most favorable plume type as far as high surface concentrations are concerned.

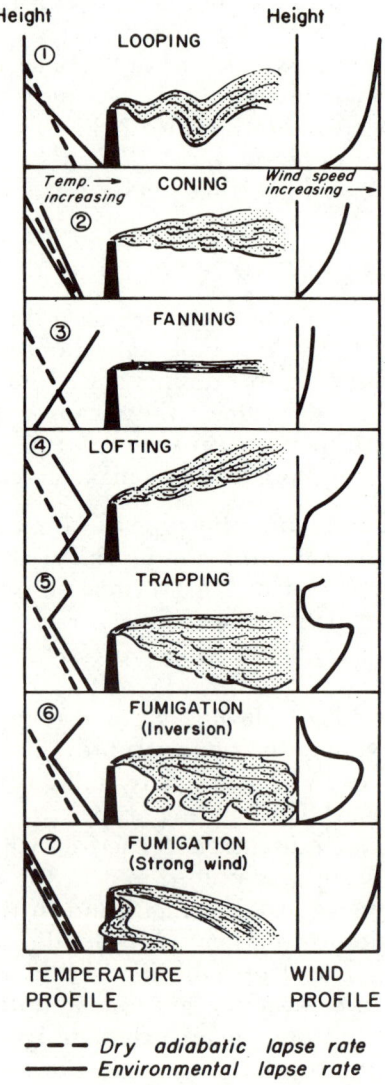

Figure 2.4 Plume behavior and meteorological conditions.

Trapping may occur at any time of the day in any season. It is associated with subsidence inversions which may persist for months, as for instance over Los Angeles, or with warm frontal inversions, which usually last less than a day. Because almost all emissions are trapped below the subsidence inversion for such an extended period of time, trapping indicates one of the worst pollution situations.

Fumigation is usually related to the time after sunrise when solar heating produces a growing unstable layer which mixes the surface air with pollutants accumulated below the nocturnal inversion. Fortunately, this type of fumigation which diffuses effluents down to the ground in large lumps usually lasts for only 30 minutes. Strong wind fumigation normally occurs under cyclonic weather with high wind speeds and strong mechanical turbulence. High concentrations of pollutants occur in the immediate vicinity of an emission source.

Calculation of pollutant concentrations. Most mathematical models for calculating pollutant concentrations from an emission source in use today are based on the statistical Gaussian approach. The main assumption for the Gaussian dispersion is that concentrations of pollutants from continuous sources (such as stacks) follow a binormal probability distribution in the vertical and crosswind direction downwind from the source.

The parameters involved in such a calculation are illustrated in Figure 2.5. A coordinate system is chosen in the downwind (x),

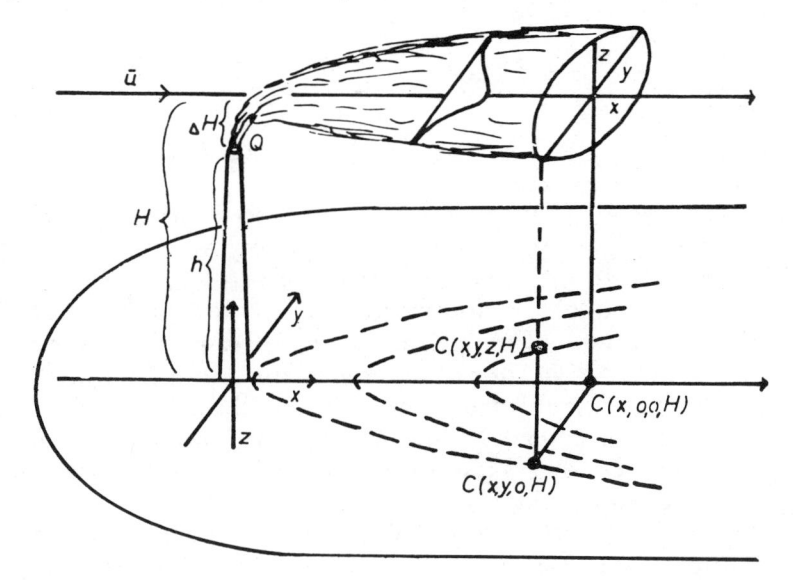

Figure 2.5 Atmospheric dispersion from an elevated source.

crosswind (y) and vertical direction (z) at the foot of the stack with height (h). The pollutant emission (Q) due to its buoyancy makes the plume rise a certain distance above the stack (ΔH). Plume rise (ΔH) plus stack height (h) give the effective stack height (H) used in all calculations. The mean wind speed $(\bar{u})$ displaces the plume downwind along the plume centerline. The plume in Figure 2.5 is projected onto a plane to show the positions for which some of the pollutant concentrations can be calculated. Suppose you want to know the ground-level pollutant concentration directly downwind; then the modified Gaussian diffusion model becomes

$$C(x,0,0,H) = \frac{Q}{\pi \bar{u} \sigma_y \sigma_z} \exp\left[-\frac{H^2}{2\sigma_z^2}\right]$$

Or the ground-level concentration at a certain distance (y) off the centerline is of interest, then

$$C(x,y,0,H) = \frac{Q}{\pi \bar{u} \sigma_y \sigma_z} \exp\left[-\frac{y^2}{2\sigma_y^2} - \frac{H^2}{2\sigma_z^2}\right]$$

Or someone lives in a high-rise apartment house and wants to know how much pollution he is subjected to from a certain stack, then

$$C(x,y,z,H) = \frac{Q}{2\pi \bar{u} \sigma_y \sigma_z}\left[\exp-\frac{y^2}{2\sigma_y^2}\right]\left[\exp-\frac{(z-H)^2}{2\sigma_z^2} + \exp-\frac{(z+H)^2}{2\sigma_z^2}\right]$$

The crosswind (σ_y) and vertical (σ_z) standard deviations of the plume are given in meters, so are the distances (y) and (z) and the effective stack height (H); the mean wind speed $(\bar{u})$ is in m/sec; the pollutant emission rate (Q) is in g/sec, and the pollutant concentration (C) is in g or $\mu g/m^3$. The important meteorological parameters in the dispersion models are the mean wind speed $(\bar{u})$ and indirectly the standard deviations of the plume (σ_y, σ_z) because the latter are a function of stability [6]. The pollutant concentration (C) is inversely proportional to $\bar{u}, \sigma_y$, and σ_z, which logically means the greater the air motion and the spread of the plume, the lower becomes the concentration. The other factors involved relate to receptor geometry (y, z) and engineering data (Q, H).

The accurate calculation of the plume rise (ΔH) which is essential for the estimation of the effective stack height (H) remains somewhat of a problem. A large number of formulae have been developed which relate the plume rise to meteorological and nonmeteorological variables [7]. Some of the difficulties which arise when devising a plume rise formula are depicted in Figure 2.6. If, for example, the stack exit velocity of the plume is much less than the wind speed,

STACK EXIT VELOCITY
MUCH LESS THAN
WIND SPEED

STACK EXIT VELOCITY HIGHER THAN WIND SPEED

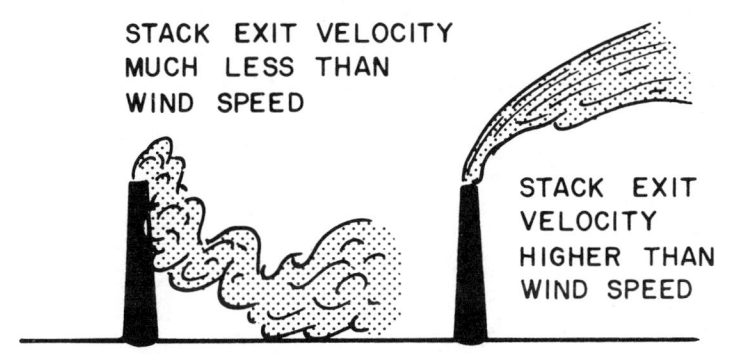

Fig. 2.6 Plume rise.

then the plume subsides in the immediate vicinity of the stack (aerodynamic downwash). From a standpoint of goodness of fit of observed and calculated plume rises, the Moses-Carson formula is probably the best at the present state of the art [8]. For all stability classes

$$\Delta H = A \left[-0.029 \frac{V_s d}{\bar{u}} + 5.35 \frac{Q_h^{1/2}}{\bar{u}} \right]$$

Corrections for stability are made by using the A-factors of 2.65, 1.08, and 0.68 for unstable, neutral, and stable conditions respectively. The mean wind speed ($\bar{u}$) has to be recorded. The stack exit velocity (V_s in m/sec), the stack diameter (d in m), and the heat emission rate (Q_h in kg cal/sec) have to be obtained from the stack operating engineer. A comprehensive survey on plume rise estimates has recently been given by Thomas, et al. [9]. For more detailed explanations of pollutant concentration calculations consult the work-and-guide books issued by the U.S. Public Health Service [10, 11], the American Society of Chemical Engineers [12], and the U.S. Atomic Energy Commission [13].

Forcasting air pollution potential. Once air pollutants are emitted into the atmosphere their subsequent fate is entirely dependent on weather processes. It would therefore be of great importance if high accumulations of pollutants could be forecast so that precautions on pending air pollution disasters could be taken. Routine forecasts of air pollution potential (FAPP) for the area east of 105° W longitude began in August 1960 [14]. By October 1963 all of the contiguous

26   ATMOSPHERIC POLLUTION

United States was included in an FAPP. The number of days of FAPP in a single year are shown in Figure 2.7. As expected the areas such as California and Appalachia with a high frequency of stagnating anticyclones (high pressure systems) and inversions are also areas with the highest frequency of FAPP. In order to interpret Figure 2.7 and understand the concept of FAPP correctly, one must

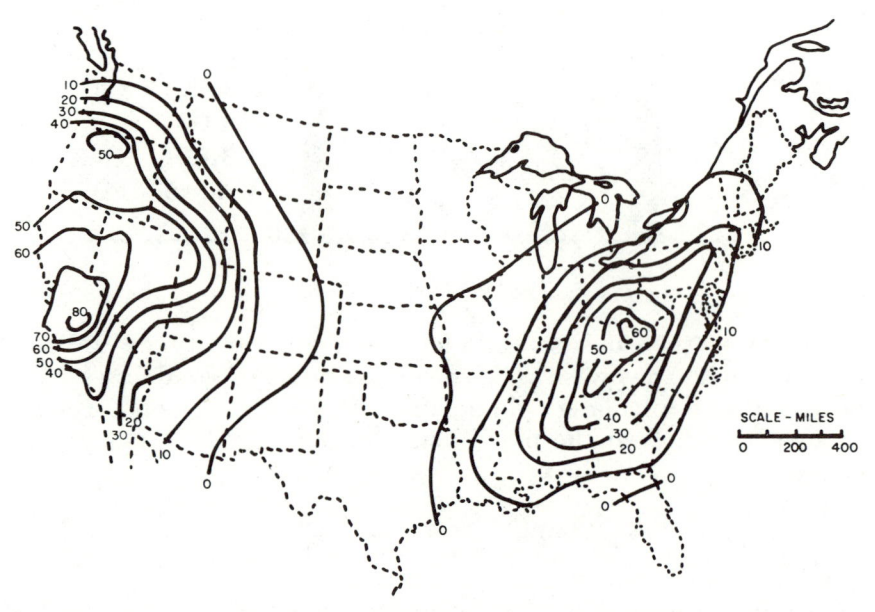

Figure 2.7   Forecast high pollution potential days. (Source: M. Smith, ed. [12])

realize that the forecasts are of pollution potential, and not of actual pollution occurrence. This means that the meteorological conditions are conducive to high air pollution concentrations irrespective of whether or not pollution sources exist in the area in question.

Two of the major FAPP criteria currently in use are the mixing depth and the average wind speed through the mixing depth (Figure 2.8). The maximum mixing depth, which usually occurs in the early afternoon, is obtained by drawing the dry adiabatic lapse rate through the maximum temperature at the surface and the environmental lapse rate [15]. The minimum mixing depth, which normally occurs before sunrise, is obtained by adding 5° C to the minimum surface temperature and following the same procedure as above. Since the temperature readings are usually taken at the airport, and since the city is on the average 5° C warmer (heat-island effect), 5° C are added to obtain a representative mixing depth for the city. The depth of the mixing layer times the average wind speed within it gives the rate of ventilation. Obviously, the greater the ventilation, the more pollutants are dispersed and diluted, and the smaller is the pollution hazard.

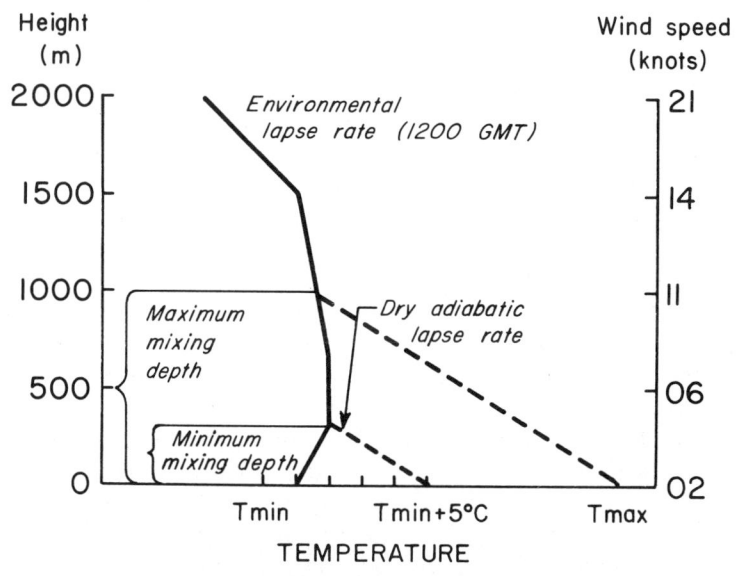

Figure 2.8 Mixing depth concept.

An air pollution forecasting system is, however, only meaningful if an alert can be issued based on legal authority which forces major pollutors to cut or shut down during the critical period. New York and Philadelphia, for example, have such legal authority. The Ruhr district in Germany is another major polluted area that has such an authoritative alert system. When the weather bureau forecasts an APP of 1,000 $\mu g/m^3$ of SO_2, an air pollution supervising agency has the authority to make power plants and industrial firms switch to a fuel with a low sulfur content, or even to shut them down. During such alert periods, incinerators may be shut down, and cities may be declared off-limits to private vehicular traffic. Only ambulances, police cars, and public transportation are then allowed in the streets. The supervising agency has thus an emergency plan which is put into action as the ambient air deteriorates. An air pollution forecasting system without legal authority to curb pollutant emissions can only suggest that people stay indoors.

SUMMARY

Once pollutants are emitted into the atmosphere, their subsequent fate is solely a function of the prevailing weather conditions. Most of the pollutants are emitted in enormous quantities into the relatively small airsheds over urban and industrial areas. The atmosphere has a great, but not an unlimited, ability to dilute and eventually remove pollutants. Without the natural removal, scavenging, diluting, and dispersing processes in the troposphere, life would perhaps no longer be possible on this earth.

The energy for all these life-sustaining weather processes is supplied by the sun. However, the lowest few thousand meters of a city's polluted atmosphere are capable of depleting the incoming solar energy by more than 60 percent. Thus, less energy is available for the major meteorological processes responsible for either high or low pollution concentrations, namely, atmospheric stability and turbulent air flow. The influence of stability and wind conditions on the dispersion of pollutants is demonstrated for the major types of plume behavior such as looping, coning, fanning, lofting, trapping, and fumigation.

The Gaussian diffusion model, which is presently the most commonly used approach to calculate pollution concentrations, is presented. The difficulty in calculating the effective emission height is pointed out. The principle of forecasting high air pollution potential within air pollution episodes is explained.

The Influence of Air Pollution on Weather and Climate

There is hardly any weather element that is not influenced by atmospheric pollution. The influence can best be demonstrated by comparing weather data obtained under "polluted" and "clean" conditions. This is most easily achieved by analysing data obtained in a city and in its rural environs.

THE EFFECT OF AIR POLLUTION ON LOCAL WEATHER AND CLIMATE

Solar radiation. A four-year comparison of solar radiation recorded at downtown Boston and Blue Hill Observatory, 10 miles outside the city center, revealed a 15 percent smaller solar radiation receipt for Boston [16]. Atmospheric pollutants can considerably reduce short-wave solar radiation as is demonstrated by comparing the reduction on a polluted and on a clean day (Figure 2.9). The shorter the wavelength the greater is the reduction. On the average, polluted city atmospheres receive 10 to 20 percent less solar radiation than surrounding rural areas. The ultraviolet (UV) radiation is 5 percent less in summer and even 30 percent less in winter [17]. A deficiency in UV radiation causes the notorious disease rickets, which was once the most common disease in the Smoky Midlands of England. Bowlegged and pigeon-breasted people are the victims of an inadequate balance of calcium, phosphorous, and vitamin D. Vitamin D-fortified milk, cod liver oil, and, of course, an unpolluted sky can control this disease.

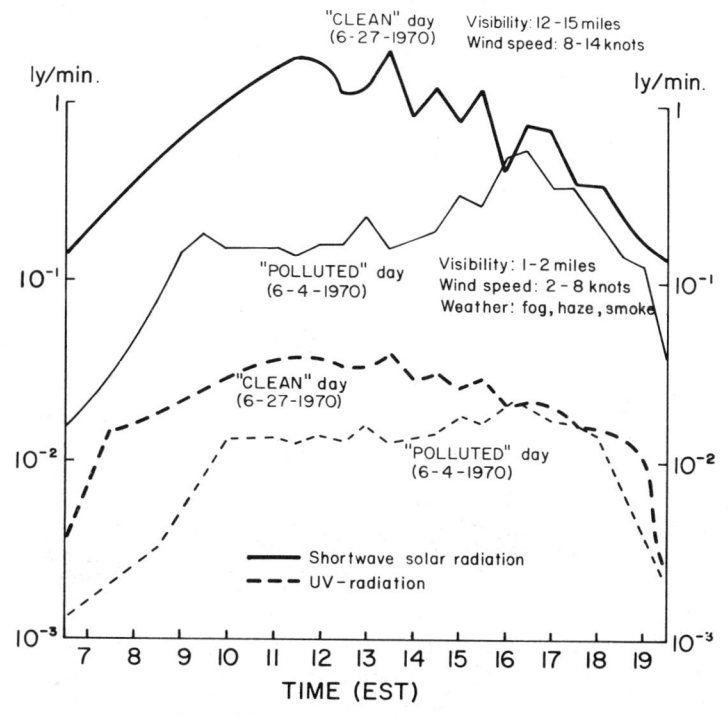

Figure 2.9 Diurnal variation of shortwave and ultraviolet solar radiation on a "polluted" and on a "clean" day in downtown Cincinnati.

Visibility and fog. A traveler approaching a city either by air or by car can often smell rather than see it. A brownish smoke pall and dramatically reduced visibility are a certain indication of human activity. Visibility, a routinely observed meteorological element, is often used in air pollution control as a rough indicator of pollution levels. Using light scattering methods, we found for Cincinnati that observed visibilities of about 9, 5, and 3.5 miles correspond to a particulate pollution concentration of 100, 200, and 300 $\mu g/m^3$ respectively. One of the most elaborate studies on visibility trends by months is that conducted by Holzworth for 28 locations distributed over the contiguous United States [18]. His results show that visibility has improved in most cities over the past 15 to 20 years. This is most likely due to a decrease in smoke pollution. Results like these and casual observations induce uncritical commentators to point for example at such apparently cleaned-up areas as Pittsburgh and London, whereas the battle against the more deleterious invisible gaseous pollution has hardly begun.

Fog is produced when there is moisture and an abundance of condensation nuclei in the air. Both are amply supplied by cities through evaporation and combustion processes. How human activity, as expressed in the amount of pollutant concentrations, affects the number of fog days on different days of the week is demonstrated for Sheffield, England, in Table 2.1. On the average there is in winter a 100 percent and in summer a 30 percent higher fog frequency in cities than in rural environs [17].

Table 2.1 Weekly variation of total number of fog days and mean smoke and SO_2 concentration ($\mu g/m^3$) in Sheffield, England, 1942-1961

	Monday	Tuesday	Wednesday	Thursday	Friday	Saturday	Sunday
Fog days	127	141	136	140	129	115	97
Smoke	493	502	490	521	527	498	476
SO_2	480	473	486	491	561	457	437

Cloudiness and humidity. Increased cloudiness in cities may be partly due to increased fog formation (a fog is a low-lying cloud), and it may partly be due to increased convection (heat-island effect) or cumulus cloud formation over cooling towers. On the average the city experiences 5 to 10 percent more cloudiness than the countryside [17].

Table 2.2 Temperature (°F) and relative humidity (%) contrasts between city center and rural environs obtained from 33 traverses in Sheffield, England

Setting	TEMPERATURE (°F)						RELATIVE HUMIDITY (%)					
	Summer		Autumn		Winter		Summer		Autumn		Winter	
	Day	Night	Day	Night	Day	Night	Day	Night	Day	Night	Day	Night
City center	70.7	55.6	63.4	54.1	32.6	26.8	34	63	40	57	69	77
Rural environs	65.3	42.8	59.6	42.6	22.7	12.2	47	95	47	86	77	96
Contrast	5.4	12.8	3.8	11.5	9.9	14.6	13	32	7	29	8	19

Industrial processes and automobiles, the largest artificial sources of moisture in cities, emit large amounts of water vapor into the air. Yet the city is drier than its surroundings. This is partly due to higher temperatures in cities which can hold relatively more moisture. Also, after a rain shower, rapid runoff of precipitation into the sewer system does not allow for much moisture evaporation. On the average, during the day the relative humidity is between 7 and 13 percent and during the night between 19 and 32 percent lower in the city center of Sheffield as compared with its rural environs (Table 2.2).

Precipitation. One of the most varying weather elements over urban and nonurban areas is precipitation. An increase in precipitation caused by cities seems to be established. The four major factors for this increase are 1. the surplus of condensation nuclei from pollutants; 2. the disturbance of the general air flow through high-rising city structures; 3. the thermal updrafts over the city heat island; and 4. the addition of water vapor from combustion processes.

One of the most widely publicized weather modifications due to air pollution is the La Porte, Indiana, precipitation anomaly. Since 1925 the year-to-year fluctuations in the annual precipitation at La Porte agree with the temporal distribution of steel production in the Chicago/Gary area [19] (Figure 2.10). The precipitation at La Porte, 30 miles downwind of the Chicago urban complex, shows a close relationship to the smoke-haze days observed in Chicago. Valparaiso,

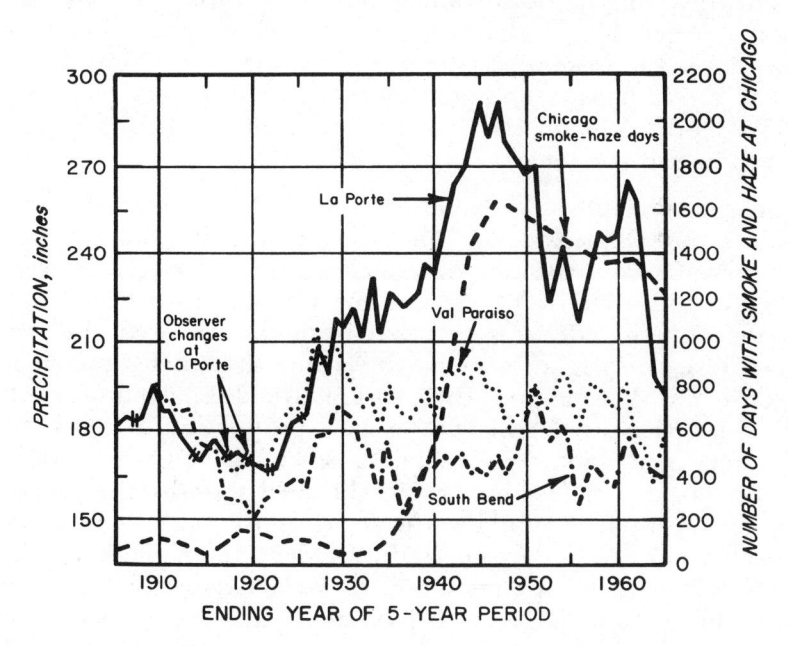

Figure 2.10 Precipitation values at selected Indiana stations and smoke-haze days at Chicago, both plotted as five-year moving totals.
(Source: S. A. Changnon, Jr., 1968 [19])

20 miles to the southwest, and South Bend, 30 miles to the northeast of La Porte, are apparently outside the narrow pollution plume, since their precipitation patterns appear not to have changed. During the 1951-1965 period La Porte had 31 percent more precipitation, 38 percent more thunderstorms, and 246 percent more hail days than surrounding locations. Changnon concluded that the differences were real and due to high emissions of ice nuclei and heat from the nearby industrial sources. Steel mills seem to be good sources for ice nuclei, which are essential in the precipitation forming processes. Also ice crystals from automobile exhausts in downwind locations of the United States cities and in laboratory experiments have been found to enhance precipitation [20].

However, if the supply of condensation nuclei is excessive, then rainfall may be reduced. This has been shown for Queensland, Australia, where excessive amounts of smoke from the burning of sugar cane decreased rainfall by 25 percent over the past 50 years [21]. Apparently there is insufficient moisture and an oversupply of muclei which can produce clouds but no rainfall.

Snowfall also seems to be triggered off more easily over a polluted urban complex. For example, the first and the last snowfall in Munich, Germany, occurs about nine days earlier and lasts about ten days longer than in the environs. Cities may have a longer snow season and in fact receive greater quantities than their rural surroundings, but the actual snow amount on the ground is much less, because it melts more easily in the warmer cities, and it is usually quickly removed by salt sprayings.

Temperature, stability, ventilation. Dust dome, haze layer, and cloudiness intercept a greater amount of solar radiation over the city than over the countryside. A greater amount of heat, however, is retained within the city because of the greenhouse effect (longwave radiation is radiated back to earth from the polluted and cloudy skies). Thus, cities tend to be much warmer than their rural environs as is shown for Washington, D.C., and Sheffield, England (Table 2.3). From some viewpoints the greater heat production in cities can be considered beneficial. For example, since the city is warmer, it requires 10 percent less heating in the cold months than the rural environs [17]. The major beneficial effect of the city heat comes into play when a whole region stifles under a thick blanket of smog. Then the city heat often initiates its own circulation system, the country breeze and convectional (vertical) air flow, which to some extent ventilate the city and dilute the pollutants. That city heat modifies the urban stability conditions has been shown for the San Francisco Bay area [22], Louisville [23], and Cincinnati [24]. Greenhouse and heat-island effects can be quite adverse factors in summer when they produce oppressive and muggy conditions. Apart from the discomfort, they also cause high electricity bills for the greater usage of air conditioning.

Table 2.3 Temperature variations (°F) with urban land use in Washington, D.C., and Sheffield, England, on a clear summer day

Land use	Washington, D. C. (E.S.T.)		Sheffield, England (G.M.T.)	
	1320	2200	1459	2337
Business center	97	85	69	54
Industry	—	—	71	56
Dense residential	96	83	70	53
Park near city center	95	84	66	46
Park near city fringe	94	78	—	—
Suburban residential	95	79	69	51
Rural environs	95	76	65	42
Contrast	3	9	6	12

Source: H. Landsberg, 1962 (for Washington, D.C.) [17].

It has been shown that air pollutants affect weather, as exemplified by hour-to-hour and day-to-day variations, and also climate, as shown by long-term modifications. On a local scale these influences can be experienced by everyone. On a global scale and extended over long periods of time the adverse effects of man-made pollution are disguised. However, because of possible catastrophic implications the long-term effects need all our attention and concern.

THE EFFECT OF AIR POLLUTION ON GLOBAL CLIMATE

The previous paragraphs have shown that air pollutants can change weather and climate considerably on a local and regional scale. The question is now, Is there sufficient evidence to assume that air pollution affects weather and climate also on a global scale?

The facts. The facts as they are known today are summarized in Figure. 2.11.

1. From the 1880s to the 1940s the mean world temperature has risen by about 0.7° F.

2. Since the 1940s the mean annual temperature of the world has decreased by 1/3 to 1/2 of a degree Fahrenheit. At first glance this seems to be insignificantly small until one realizes that the last ice age was brought about by a temperature drop of only 4-5° F.

3. Since the 1880s and continuing until today there has been an increased sunspot activity. This would mean an increase in solar

34 ATMOSPHERIC POLLUTION

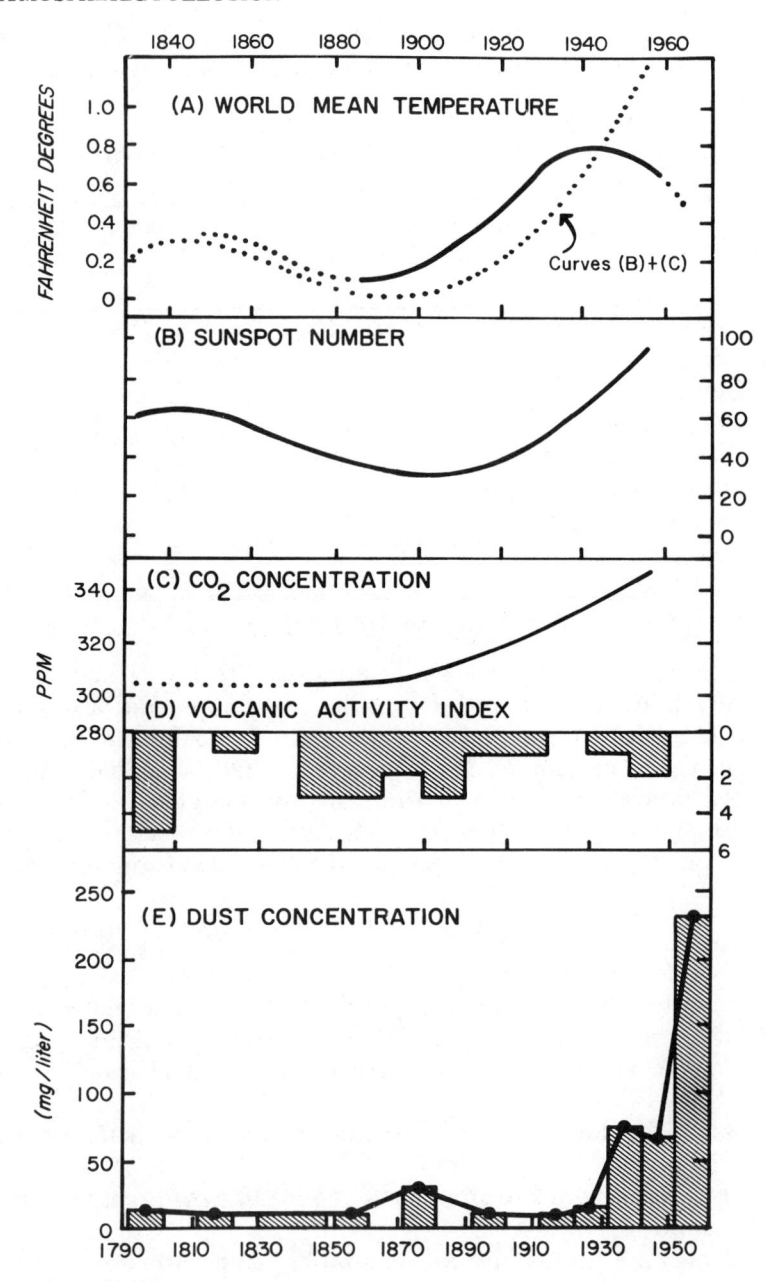

Figure 2.11 Trends of mean world temperature, sunspot number, carbon dioxide, volcanic activity, and dustfall in the Caucasus. (Sources: J. M. Mitchell, Jr., 1961 [25]; F. F. Davitaya, 1969 [26]; W. M. Wendland and R. A. Bryson, 1970 [27])

energy output, which in turn would mean an increase in world temperatures. However, the mean world temperature has been decreasing since the 1940s, while the sunspot activity is still increasing. Consequently there must be a more influential factor at work.

4. The carbon dioxide (CO_2) content of the atmosphere has been increasing since the 1880s with increasing industrialization. By preventing longwave or heat radiation from leaving the earth, and thus acting like a greenhouse, the increased CO_2 content in the atmosphere should also cause an increase in mean world temperatures. However, since the 1940s the mean world temperature has been falling, while the CO_2 content of the atmosphere continues to rise. Consequently, a more influential factor must be in operation.

5. Increased volcanic activity through scattering solar energy back into space would decrease the mean global temperature. Since the violent eruptions of Krakatoa in 1883, Santa Maria and Pelee in 1909, and Katmai in 1912 [26], there have been a few major eruptions in more recent times, such as Mt. Spurr in Alaska in 1953 and Mt. Agung on Bali in 1963 [28]. But in general, volcanic activity has decreased which should have favored a general increase in mean world temperature. Consequently, a more influential factor must be in operation.

6. Another factor is dust and fine suspended particulate matter in the atmosphere. Dust concentration in Figure 2.11 shows an enormous increase since the 1930s with a slight stagnation period during World War II. Since no increased volcanic activity is reported during this time period, the sharp rise must be attributed to man-made pollution. An increase in the dust loading of the atmosphere would indeed lower the mean world temperatures.

The important question is now, What changes in the above elements are large enough to trigger a change in one direction or the other. Using Bryson's heat budget approach the various theories of climatic change can be related [29]. The energy received from the sun and the heat lost to space by the earth can be expressed by

$$ScA = KeT^4 \quad (4c)$$

where S is the incoming solar radiation received at the earth, c is the cross-sectional area of the earth, A is the amount of radiation absorbed, K is a constant, e is the effective emissivity of the earth, and T is the average global temperature in degrees Kelvin. What this formula is in fact saying is that the incoming solar energy is only intercepted by a cross-section (c) of the earth, but that it is lost as heat radiation over the whole globe ($4c$). The above formula can be

36   ATMOSPHERIC POLLUTION

solved for the mean temperature of the earth T. The effect on T through a change in solar energy S, absorptivity A, and effective emissivity e can be discussed.

Solar radiation. The solar energy reaching the earth's atmosphere is about 2 ly/min (2 cal/cm² /min). About 60 to 65 percent of this incoming radiation is absorbed and eventually radiated back into space. Since of the solar energy of 2 cal/cm² /min, 60 percent is reemitted from the total surface of the earth, i.e., 2 cal/cm² /min x 0.60/4, each square centimeter of the earth loses about 0.30 cal/min by heat radiation. A 1 percent decrease in solar radiation would decrease the mean temperature of the earth by about 0.8° C (1.4° F) [29].

Reflectivity. A decrease in solar intensity can be produced by an increase in absorptivity. In meteorology the term reflectivity or albedo, which is the fraction of solar energy that is reflected, is used rather than the term absorptivity, which is the amount of radiation absorbed. The albedo of the earth, which is about 39 percent, depends, for instance, on the amounts of clouds and the dust content in the atmosphere, on the amount of snow and ice on the ground, and on the type of ground cover (forests or deserts, etc.). Increased cloudiness, such as produced over urban and industrial areas or by contrails of jet planes, suspended particulate pollutants from human and volcanic activity, or increased snow and ice covers over polar and alpine regions, increases the earth's reflectivity and thus decreases the temperature.

The amount of dust found in firn layers of the Caucasus showed hardly any variation from 1790 to 1930 [26]. But between the short time period of 1930 to 1963 the amount of dust deposited in this remote area increased 19-fold (Figure 2.11). Davitaya attributes this enormous increase to intensification of human activity.

Table 2.4 Turbidity variations in Washington, D.C., and Davos, Switzerland

Years	Mean annual turbidity	Percent increase	Number of Aerosol particles (cm^{-2})	Increase in (mill./cm²)
		Washington, D.C.		
1903-1907	0.098		0.54 x 10⁶	
1962-1966	0.154	57	0.82 x 10⁸	28
		Davos, Switzerland		
1914-1926	0.024		0.125 x 10⁸	
1957-1959	0.043	88	0.230 x 10⁸	10.5

Source: R.A. McCormick and J. H. Ludwig, 1967 [30].

McCormick and Ludwig report that over a 60-year period the turbidity or haziness over Washington, D.C., increased by 57 percent, and in 30 years over Davos, Switzerland, by 88 percent (Table 2.4) [30]. Translated into numbers of particles in the air this means that there are now in Washington's air 28 million more aerosols (small solid and liquid particles) in the radius range of 0.1 to 1.0 μm per square centimeter than there were 60 years ago. McCormick and Ludwig attribute two-thirds of Washington's increased turbidity to man-made local pollution.

The increase in global turbidity is further demonstrated by the Mauna Loa data of Hawaii (Figure 2.12). The overall trend shows

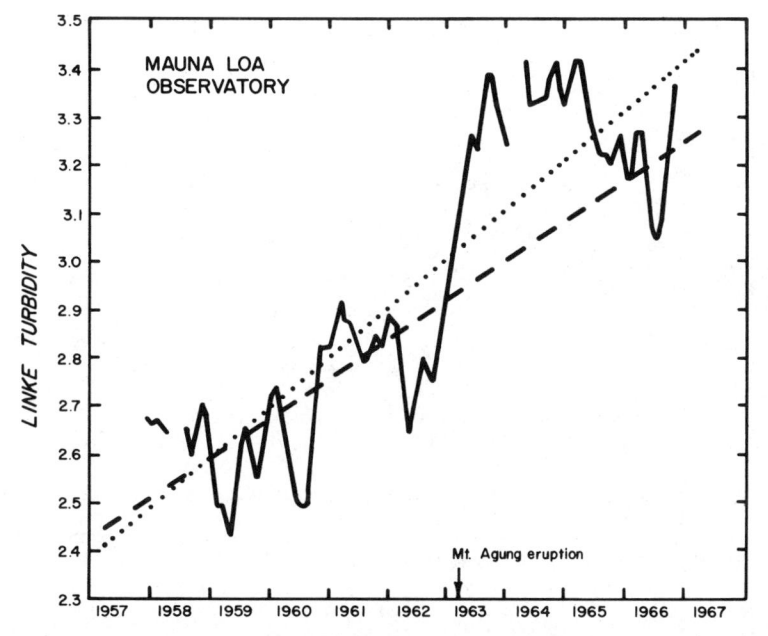

Figure 2.12 Increasing turbidity at Mauna Loa, Hawaii. (Souce: R. A. Bryson and J. E. Kutzbach, 1968 [28])

quite decisively an increase in background turbidity, since Mauna Loa is remote from any man-made pollution. The dotted line shows the linear trend for the whole period, whereas the dashed line represents the turbidity trend without the Mt. Agung volcanic eruption. Even without the interference of Mt. Agung, the turbidity increase is still about 30 percent between 1957 and 1967 [31]. A 1 percent increase in the reflectivity of the earth (39 percent) would decrease the mean global temperature by about 1.7° C (3.1° F) [29].

Effective emissivity. A decrease in the emission of heat radiation will result in an increase of temperature. This increase can be due to an

38 ATMOSPHERIC POLLUTION

increase in CO_2, clouds, ozone, water vapor, or dust. Carbon dioxide is one of the major factors in preventing heat from escaping into space (greenhouse effect).

Since 1860, CO_2 from burning of fossil fuels has increased by 15 percent [32]. If all the known fossil fuel reserves were burned, this would increase the present CO_2 content of the atmosphere 17 times [33]. The present CO_2 increase is a worldwide phenomenon as can be seen by the CO_2 trends for Mauna Loa, Hawaii, and the Antarctic (Figure 2.13). The average CO_2 increase at Mauna Loa is 0.06 ppm per month or 0.72 ppm per year, and the data for the Antarctic are 0.1 ppm and 1.2 ppm respectively [33].

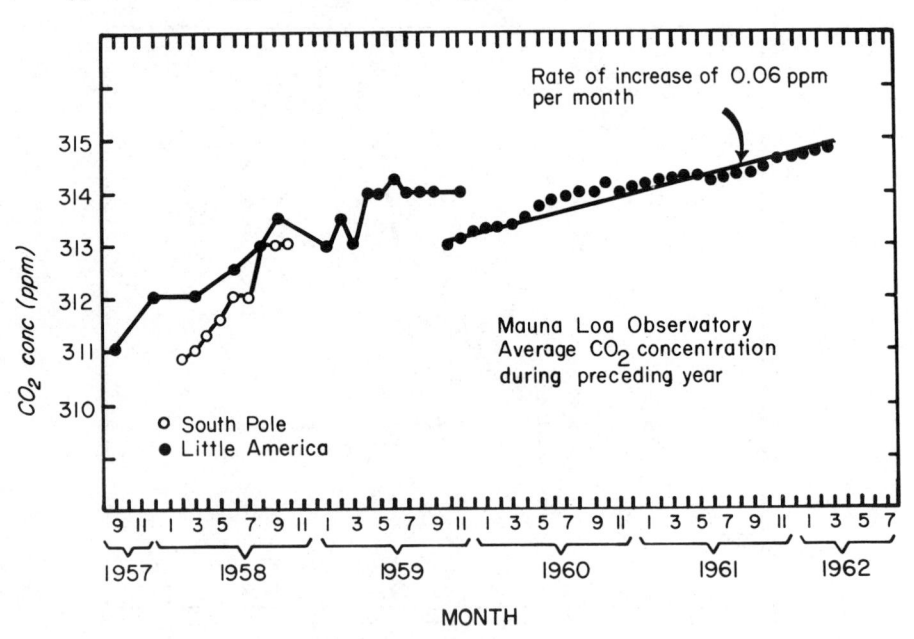

Figure 2.13 Variation of CO_2 concentrations at Mauna Loa, Hawaii, and in the Antarctic. (Sources: G. N. Plass, 1957 [33]; J. C. Pales and C. D. Keeling, 1965 [34])

Although the burning of fossil fuels is the major factor in increasing the CO_2 and reducing the O_2 content, cement and limestone production, and above all, the destruction of the vegetation cover caused by man's activities are important influences. It has been estimated that vegetation uses about 150 million tons of CO_2 in photosynthesis. An increase in forest areas would decrease the CO_2 content. However, at the 1962 rate of fuel combustion it would require about 2.7 billion acres of new forest in the United States to absorb the CO_2 produced. This would amount to 120 percent of the area of all 50 states [32].

Oceans also absorb CO_2. Since this gas is more soluble at lower temperatures, it is being absorbed in polar regions and released over tropical waters. However, the overall amount of CO_2 that is absorbed by the oceans is unknown.

The effect of CO_2 on temperature changes is quite complicated as the following reasoning by Peterson shows [32]: Increased CO_2 production increases world temperatures, which in turn increase evaporation and the water vapor content in the atmosphere. This could on the one hand increase the temperatures (greenhouse effect), but on the other hand, it could decrease the temperature through increased cloudiness and thus greater reflectivity. This explains in part the apparently contradicting statements that we are approaching an ice age or a heat age. Since CO_2 is only one part of the story in the greenhouse effect, it is better to argue in terms of effective emissivity, i.e., the amount of heat that in effect leaves the earth. By increasing the effective emissivity by 1 percent from 55 percent to 56 percent, the mean global temperature would decrease by about 1.2° C (2.2° F). The temperature change would be much smaller if the CO_2 effect were considered alone [29].

By now the reader will be wondering which of the factors would most likely produce the observed changes and what can we possibly expect in the future. Figure 2.11 has shown that the mean annual temperature of the globe increased by 0.7° F between the 1880s and the 1940s. Bryson has calculated that this could be due to an increase in solar radiation reaching the earth of about 0.5 percent, or a decrease of albedo of about 0.25 percent, or a decrease of effective emissivity of 0.3 percent. Unfortunately, none of these factors can be assessed with sufficient accuracy. However, as Figure 2.11 has shown, the tremendous increase in dust pollution is the only plausible cause for the recent temperature decrease. Thus, some experts believe that the overriding factor causing the decrease in mean world temperature is the effect global air pollution has on reflecting solar radiation back to space [29]. It is reasonable to assume that temperatures will continue to decrease, if the emission of pollutants into the atmosphere continues at the present global scale.

SUMMARY

The influence of air pollution on weather and climate is demonstrated by comparing weather data recorded in polluted cities with those obtained in relatively clean rural surroundings. On the average, polluted city atmospheres receive 10 to 20 percent less solar radiation than surrounding rural areas. Visibility seems to have been improved over the past 20 years, presumably due to a decrease in particulate pollution. In winter, fog is about 100 percent more frequent and in summer 30 percent more frequent in cities than in rural environs. Cities experience on the average a 5 to 10 percent increase in cloudiness as compared to the countryside. La Porte, in the lee of the Chicago pollution plume, had 31 percent more precipitation, 38 percent more thunderstorms, and 246 percent more hail days than neighboring areas. Cities are up to 15° C warmer than surrounding areas. This heat-island effect makes city air more unstable and produces a circulation system called the country breeze.

On a global scale the effects of pollution on modifying the climate are less obvious. It is shown that from the 1880s to the 1940s the mean world temperature has increased by about 0.7° F. Since the 1940s the mean world temperature has decreased by about 0.5° F. Also since the 1880s until today the sunspot activity and the CO_2 content of the atmosphere have increased. Volcanic activity does not seem to have increased but some experts have found an increase in background particulate pollution levels.

Discussing a potential climatic change in terms of solar radiation, reflectivity, and effective emissivity, it has been argued that the observed decrease in mean world temperature might to some extent be due to back-scattering of solar radiation from the dust loading of the atmosphere. Most recent investigations of the Mauna Loa, Hawaii, turbidity data seem, however, to indicate that the increase in background dust loading might only have been temporary.

REFERENCES CITED

[1] P. A. Leighton, "Geographical Aspects of Air Pollution," *Geogr. Rev.* 56(2), 151-174, Apr., 1966.
[2] R. E. Munn, "Air Pollution Meteorology," *Occup. Health Rev.* 20(3-4), 1-8, 1968-1969.
[3] M. Neiburger, "Meteorological Aspects of Air Pollution," *Arch. Environ. Health* 14, 41-45, Jan., 1967.
[4] J. J. Fuguay, "Natural Removal Processes in the Atmosphere," in *Meteorological Aspects of Air Pollution*, USDHEW, PHS, NAPCA, Res. Triangle Park, N. C., Feb., 1969.
[5] S. M. Greenfield, "Rain Scavenging of Radioactive Particulate Matter from the Atmosphere," *J. of Meteor.* 14(2), 115-125, 1957.
[6] F. Pasquill, *Atmospheric Diffusion*, D. Van Nostrand Co., Inc., New York, 1962, 297 pp.
[7] H. A. Panofsky, "Air Pollution Meteorology," *Amer. Scientist* 57(2), 269-285, 1969.
[8] J. E. Carson and H. Moses, "The Validity of Several Plume Rise Formulas," *JAPCA* 19(11), 862-966, Nov., 1969.
[9] F. W. Thomas, et al., "Plume Rise Estimates for Electric Generating Stations," *JAPCA* 20(3), 170-177, Mar., 1970.
[10] USDHEW, PHS., *Meteorological Aspects of Air Pollution*, NAPCA, Res. Triangle Park, N. C., Feb., 1969.
[11] D. B. Turner, *Workbook of Atmospheric Dispersion Estimates*, USDHEW, PHS, NAPCA, Cincinnati, Ohio, revised 1969.
[12] M. Smith, ed., *Recommended Guide for the Prediction of the Dispersion of Airborne Effluents*, Am. Soc. Mech. Eng., New York, 1968.
[13] D. H. Slade, ed., *Meteorology and Atomic Energy 1968*, U.S. Atomic Energy Commission, Div. Techn. Inf., Oak Ridge, Tenn., July, 1968.
[14] M. E. Miller and L. E. Niemeyer, "Air Pollution Potential Forecasts—A Year's Experience," *JAPCA* 13(5), 205-210, 1963.
[15] C. R. Hosler, "Climatological Estimates of Diffusion Conditions in the U. S.," *Nuclear Safety* 5(2), 184-192, Winter, 1963-64.
[16] J. F. Hand, "Atmospheric Contamination over Boston, Mass.," *Bull. Am. Met. Soc.* 30(7), 252-254, 1949.
17] H. Landsberg, *Physical Climatology*, 2nd ed., Gray Printing Co., Inc., DuBois, Penn., 1962, 446 pp.
[18] G. C. Holzworth, *Some Effects of Air Pollution on Visibility in and near Cities*, Rpt. SEC TR A62-5, Taft Sanitary Engineering Center, Cincinnati, Ohio, 69-88, 1962.
[19] S. A. Changnon, Jr., "The La Porte Weather Anomaly—Fact or Fiction," *Bull. Am. Met. Soc.* 49(1), 4-11, Jan., 1968.
[20] V. J. Schaefer, "The Inadvertent Modification of the Atmosphere by Air Pollution," *Bull. Am. Met. Soc.* 50(4), 199-206, Apr., 1969.
[21] J. Warner, "A Reduction in Rainfall Associated with Smoke from Sugarcane Fires—An Inadvertent Weather Modification?" *J. Appl. Met.* 7, 247-251, 1968.
[22] F. S. Duchworth and J. S. Sandberg, "The Effect of Cities upon Horizontal and Vertical Temperature Gradients," *Bull. Am. Met. Soc.* 35(5), 198-207, 1954.
[23] G. A. DeMarrais, "Vertical Temperature Difference Observed over an Urban Area," *Bull. Am. Met. Soc.* 42(8), 548-554, 1961.

[24] J. F. Clarke, "Nocturnal Urban Boundary Layer over Cincinnati, Ohio," *Mon. Wea. Rev.* 97(8), 582-589, 1969.
[25] J. M. Mitchell, Jr., "Recent Secular Changes of Global Temperature," *Ann. N. Y. Acad. Sci.* 95(1), 235-250, 1961.
[26] F. F. Davitaya, "Atmospheric Dust Content as a Factor Affecting Glaciation and Climatic Change," *AAAG* 59(3), 552-560, Sept., 1969.
[27] W. M. Wendland and R. A. Bryson, "Atmospheric Dustiness, Man, and Climatic Change," *Biol. Conservation* 2(2), 125-128, Jan., 1970.
[28] R. A. Bryson and J. E. Kutzbach, "Air Pollution," Assoc. Am. Geogr. Resource Paper No. 2, 1968, pp. 28-31.
[29] R. A. Bryson, "All Other Factors Being Constant...," *Weatherwise* 21(2), 56-61, 94, Apr., 1968.
[30] R. A. McCormick and J. H. Ludwig, "Climate Modification by Atmospheric Aerosols," *Science* 156, 1358-1359, June, 1967.
[31] J. T. Peterson and R. A. Bryson, "Atmospheric Aerosols: Increased Concentrations During the Last Decade," *Science* 162, 120-121, Oct., 1968.
[32] E. K. Peterson, "Carbon Dioxide Affects Global Ecology," *Env. Science Technology* 3(11), 1162-1169, Nov., 1969.
[33] G. N. Plass, "The Carbon Dioxide Theory of Climatic Change," *Procdgs. Conf. Recent Res. Climatol.*, H. Craig, ed., Committee on Res. in Water Res., La Jolla, 1957, pp. 81-92.
[34] J. C. Pales and C. D. Keeling, "The Concentration of Atmospheric CO_2 in Hawaii," *J. Geophys. Res.* 70(24), 6053-6076, Dec., 1965.

CHAPTER 3

HEALTH ASPECTS OF AIR POLLUTION

Air Pollution Episodes

It is very difficult to demonstrate chronic effects caused by continued exposure to air pollution. However, air pollution episodes with extremely high concentrations have been found to cause acute sickness and death. It is therefore reasonable to presume that a prolonged exposure to small concentrations will also result in adverse effects. From a community health point of view the chronic effects may be the more important. However, it was the major air pollution disasters with their thousands of excess deaths that finally resulted in some control measures.

MAJOR AIR POLLUTION DISASTERS

The major air pollution disasters which occurred in modern times are summarized in Table 3.1. The excess death rate is estimated from death rates prior to and after the acute period.

Meuse Valley, Belgium, 1930. One of the first documented episodes in modern times which aroused worldwide interest occurred in the Meuse Valley of Belgium in December, 1930. Trapped by an inversion, pollutants accumulated in this steep-sided valley of 15 miles length. Coke ovens, steel mills, blast furnaces, zinc smelters, glass factories, and sulphuric acid plants produced an estimated SO_2 concentration of 8 ppm [7]. This corresponds to about 22,600 $\mu g/m^3$ which is well above the SO_2 value of 1,000 $\mu g/m^3$ which Waller and Commins used to delineate episodes of high pollution [8]. Within a few days more than 600 people fell ill, and 63 people died from the polluted air. Unfortunately no measurements were made. There seems, however, little doubt that the major culprit was sulphur dioxide which, with the help of fog droplets, oxidized to sulphuric acid mist with a particle size small enough to penetrate deeply into the lungs [9].

Table 3.1 Major air pollution episodes

Date	Place	Excess deaths
Feb. 1880	London, England	1,000
Dec. 1930	Meuse Valley, Belgium	63
Oct. 1948	Donora, Penn., U.S.	20
Nov. 1950	Poca Rica, Mexico	22
Dec. 1952	London, England	4,000
Nov. 1953	New York, U.S.	250
Jan. 1956	London, England	1,000
Dec. 1957	London, England	700-800
Dec. 1962	London, England	700
Jan./Feb. 1963	New York, U.S.	200-400
Nov. 1966	New York, U.S.	168

Sources: W.T. Russel, 1926 [1]; H. Heimann, 1961 [2]; L. Greenburg, et al., 1962 [3]; A. E. Martin, 1964 [4]; L. Greenburg, et al., 1967 [5]; M. Glasser, et al., 1967 [6].

Donora, Pennsylvania, 1948. Air pollution disasters are no speciality of Europe. In October 1948 the United States experienced its first pollution tragedy in the small town of Donora in the Monongahela River Valley, 20 miles southeast of Pittsburgh. Effluents from a number of industries such as a sulphuric acid plant, a steel mill, and a zinc production plant became trapped in a shallow valley inversion to produce an unbreathable mixture of fog and pollution. About 6,000 people or 43 percent of the population suffered various degrees of illnesses, such as sore throats, irritation of the eyes, nose, and respiratory tract, headaches, breathlessness, vomiting, and nausea. Instead of the 2 deaths which would have been normally expected over the three-day period, the episode claimed a toll of 20 deaths [10]. Again no ambient measurements were made during the disaster. It was suggested that sulphur dioxide reached peak values of about 5,500 $\mu g/m^3$ [9].

Poza Rica, Mexico, 1950. The disaster which struck Poza Rica, a town of 15,000 people on the Gulf of Mexico, originated from an accident at one of the local factories which recovers sulphur from natural gas. The release of hydrogen sulfide into the ambient air lasted for only 25 minutes. The spread of the gas under a shallow inversion with foggy and calm conditions killed 22 people and hospitalized 320.

Cincinnati, Ohio. 1968. A similar accident with a fortunately less tragic ending occurred on August 25, 1968, in Cincinnati. About 2,500 pounds of SO_2 escaped into the air from a burst pipe at a chemical plant located in the northern industrial part of Cincinnati. The release of SO_2 lasted from 12:20 A.M. to 8:00 A.M. Eastern Standard Time. Two lower middle-class communities, one black and one white, about 200 meters to the east of the plant were affected. People were brutally awakened in the middle of the night by a rotten-egg smell and difficulty in breathing. Fortunately nobody was killed. Many of those who consulted their doctors complaining about stomach aches were told to destroy the fruit and vegetables which grew in their gardens.

Approached for compensation, the plant sent a team of lawyers for inspection, who ate an apple here and a tomato there in order to demonstrate how harmless the poisoning was. Some experts were called in by the company to reconstruct the possible spread of the SO_2 plume. Since neither meteorological nor air pollution measurements had been made near the scene of the accident, no calculations of the spread of the plume could be made. However, by plotting the vegetation damage on a map, it was quite obvious that the SO_2 plume had covered both neighborhoods. The firm made, however, no compensation, arguing that it was the end of the season anyway, and that all the plants would come back undamaged the following year.

London, England, 1952. From December 5 to 8, 1952, London experienced the worst air pollution disaster ever reported. The meteorological conditions were ideal for a pollution buildup. Anticyclonic or high pressure weather with stagnating continental polar air masses trapped under subsidence inversions produced a shallow mixing layer with an almost complete absence of vertical and horizontal air motion. The millions of open fireplaces and industry supplied the hygroscopic condensation nuclei for the moist London air to form a dense fog. The mean daily temperatures were constantly below the 80-year average for that time of the year, ensuring an increased output of pollutants from space heating and power plants.

With such adverse dispersion conditions, pollutants were bound to reach peak concentrations. Figure 3.1 and Table 3.2 show that the highest daily smoke values reached 4.46 mg/m³ (4,460 µg/m³) and the highest daily SO_2 values 1.34 ppm (3,830 µg/m³), which is nine times and five times the concentration normal for the November-December period [11].

A comparison of the meteorological and pollution conditions in Figure 3.1 with the death rates shows clearly that elderly people

46 ATMOSPHERIC POLLUTION

were particularly affected, whereas the age group under 34 did not react adversely at all. Deaths from bronchitis increased by a factor of 10, influenza by 7, pneumonia by 5, tuberculosis by 4.5, other respiratory diseases by 6, heart diseases by 3, and lung cancer by 2. When a change in weather (frontal system) finally cleared the fog away, 4,000 Londoners had perished in their "pea soup" [11].

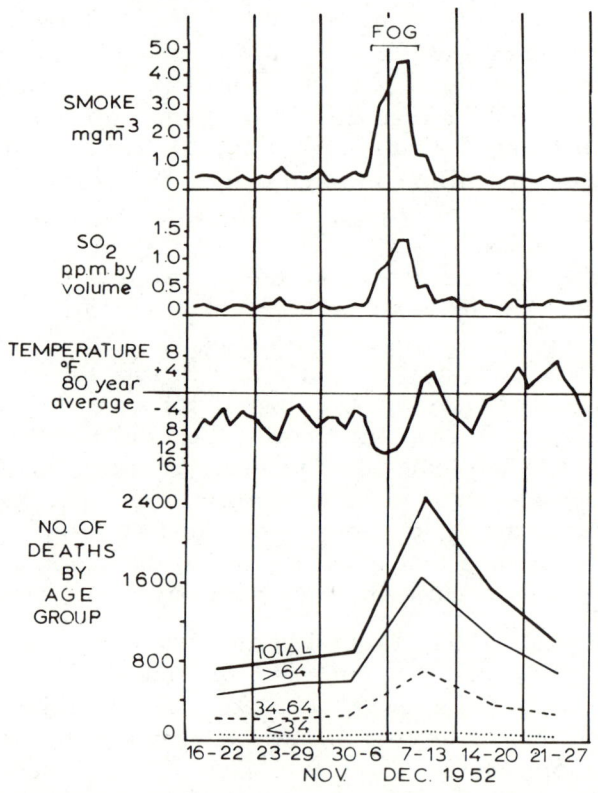

Figure 3.1 The London fog disaster, 1952. (Source: J. A. Scott, 1953 [11])

Pollution disasters with similarly high concentrations occurred in 1957-1958 and again in 1962-1963 (see Table 3.2) without the great number of casualties of 1952 (see Table 3.1). Animal tests suggest that when particulate matter and SO_2 concentrations, normally considered to be harmless, existed together as gas-aerosol combinations, they produced purple hemorrhage and paralysis of respiratory tracts. If, additionally, these gas-aerosol mixtures occurred in a super cool state such as was the case in the 1952 fog, they showed, on reaching the lungs, a higher toxicity than hydrocyanic acid (HCN).

The 1952 London smog disaster had one sinister beneficial effect: the Clean Air Act of 1956. This act mainly controls the

Table 3.2 Concentrations of smoke and sulphur dioxide during air pollution episodes in London

Winter	Date of episode		Duration of episode days	Concentrations in $\mu g/m^3$ air			
				Sulphur dioxide		Smoke	
				Highest daily	Highest hourly	Highest daily	Highest hourly
1952-53	5-9	Dec. 52	5	3830	— —	4460	— —
	20-21	Jan. 53	2	1670	— —	1480	— —
	3-4	Mar. 53	2	2179	— —	1660	— —
	24-25	Mar. 53	2	1687	— —	1370	— —
1953-54	None						
1954-55	None						
1955-56	4-6	Jan. 56	3	1430	— —	2830	9700
1956-57	18-19	Nov. 56	2	1373	— —	588	— —
1957-58	3-5	Dec. 57	3	3335	4200	2417	7200
	30-31	Jan. 58	2	1350	2430	863	1730
1958-59	28-30	Jan. 59	3	1850	4570	1723	3980
	16-19	Feb. 59	4	1584	4460	1486	2690
1959-60	12-13	Nov. 59	2	1467	3260	1280	3230
1960-61	7-9	Dec. 60	3	1338	1510	468	750
	9-10	Mar. 61	2	1164	1370	324	400
1961-62	21-22	Nov. 61	2	1052	2230	274	710
1962-63	3-7	Dec. 62	5	3834	5650	3144	4700
	23-26	Jan. 63	4	1968	3060	766	3560
	25-Feb-2	Mar. 63	6	1206	1480	311	900
1963-64	21-22	Jan. 64	2	1548	2200	702	1570
1964-65	None						
1965-66	None						

Note: Episodes of high pollution have been defined as periods of two or more consecutive days during which the daily mean concentration of smoke or sulphur dioxide exceeded 1000 $\mu g/m^3$.

Source: R. E. Waller and B. T. Commins, 1966 [8].

visible pollution which is notable in the reduced smoke concentrations in Table 3.2. The more deleterious SO_2 pollution remains almost unaffected. It would be interesting to see whether during the last four years episodes of high air pollution continued to occur with equal frequencies during the winter months.

New York, 1953, 1962-1963, 1966 [3, 12]. New York City also experiences its air pollution disasters, such as the one in 1953 and more often during the 1960s, causing hundreds of excess deaths (Table 3.1). New York with the nation's highest SO_2 concentrations often avoids air pollution disasters because of its excellent ventila-

48 ATMOSPHERIC POLLUTION

tion. If on occasions such as the December 1962 episode, adverse weather conditions prevail as given by McCarroll (see Figure 3.2) in terms of hours with low wind speeds and occurrence of shallow inversions, then the SO_2 and smoke concentrations reach peak values. Total deaths increased to 296, which was in excess of even three standard deviations above the expected mortality for that week.

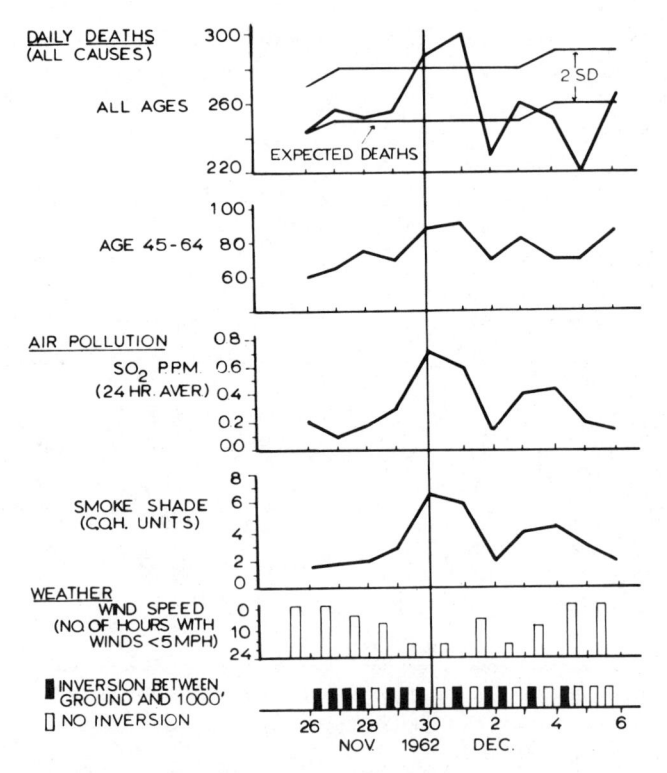

Figure 3.2 Mortality peaks in New York City compared with certain air pollution and weather recordings, November 26-December 6, 1962. (Source: J. McCarroll, 1967 [12])

WORLDWIDE AIR POLLUTION DISASTERS

Improved and extended air pollution networks have proven that air pollution disasters can be worldwide phenomena. For example, the episode starting November 27 and ending December 5, 1962, covered the whole eastern United States. The diagram of Figure 3.3 for the CAMP (Continuous Air Monitoring Program) station in downtown Cincinnati shows a plastic model of the buildup of SO_2 during the episode [13].

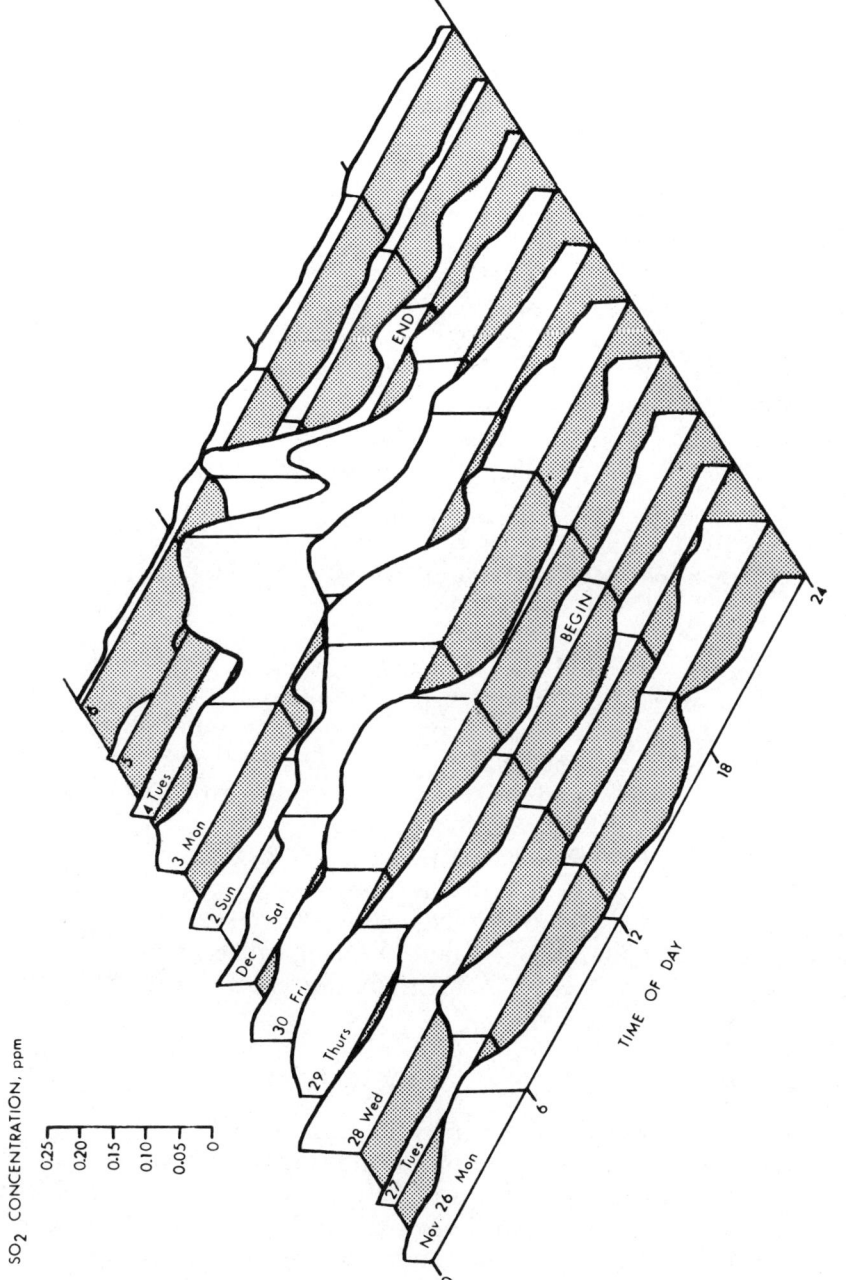

Figure 3.3 SO$_2$ levels recorded at the Cincinnati CAMP Station, November 26-December 6, 1962. (Source: D. A. Lynn, et al., 1964 [13])

In Europe, London (December 5-7) experienced its worst episode since 1952 with 700 excess deaths and increased morbidity (see Figure 3.1 and Table 3.2). In Sheffield, England (December 4-7), short-period smoke (Figure 3.4) and SO_2 pollution concentrations were also recorded. Between December 4 and 7, the smoke values were 16 times and the SO_2 values were 12 times higher than the concentrations normal for December.

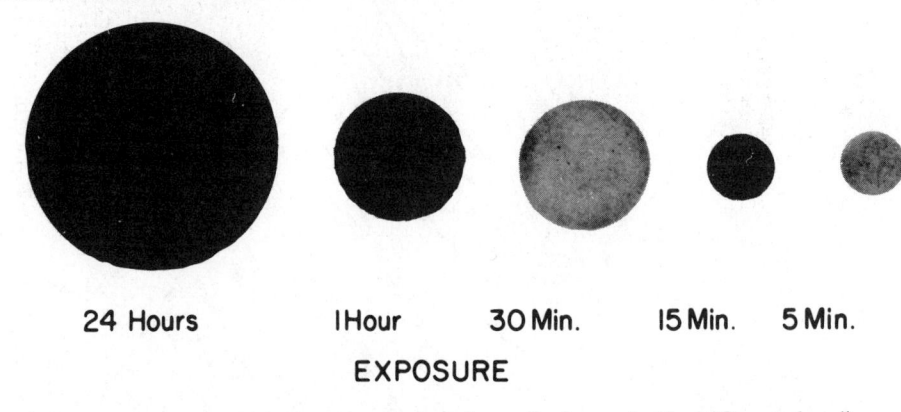

24 Hours 1 Hour 30 Min. 15 Min. 5 Min.
EXPOSURE

Figure 3.4 Smoke samples taken during the pollution episode of December 5, 1962, in Sheffield, England.

In Rotterdam, Holland (December 2-7), the SO_2 levels increased to almost five times the normal level. Increased pollution levels were also recorded in Paris, the Ruhr district of Germany, Frankfurt, and Prague. Hamburg (December 3-7) reported SO_2 levels five times, and dust levels twice the usual levels. Mortality due to heart disease increased [9].

In Asia, Osaka (December 7-10) claimed 60 excess deaths due to high pollution levels [9]. It is interesting to note that this episode started east of the Mississippi on November 27, 1962, and bringing disaster around half the globe, ended in Japan on December 10. The episode was neither recorded in Australia nor on the West Coast of the United States.

LESSONS FROM AIR POLLUTION DISASTERS

All disasters occurred in the winter months in the northern temperate zone with dense populations and heavy industrialization. Adverse meteorological conditions such as stagnating air masses under shallow inversions with impeded ventilation and dispersion played a decisive role. Not one pollutant, but combinations of two or more gas-aerosol mixtures in moist and cold weather caused synergistic (highly enhanced) damage to health. Air pollution disasters at the time of their occurrence have never been fully appreciated.

Therefore, guidelines for protection were never issued. Air pollution disasters have proven that pollutants in certain combinations and concentrations are detrimental to health and often lethal.

SUMMARY

Major air pollution disasters resulting in hundreds and thousands of excess deaths have periodically occurred since the Industrial Revolution. Air pollution episodes are not restricted to major centers of population and heavy industrialization. Pollution sources entrenched in steep-sided valleys, for example, can also lead to air pollution disasters if the pollutants become trapped in a stagnating air mass under a shallow inversion lid.

Major air pollution episodes of recent times have occurred in the Meuse Valley, Belgium; Donora, Pennsylvania; London, and New York. Poza Rica, Mexico, and Cincinnati, Ohio are examples of so-called industrial pollution accidents. Some of the air pollution episodes seem to occur now on a continental and sometimes hemispherical scale.

Air pollution episodes with their excessively high air pollutant concentrations and their synergistic effects have proven exceedingly detrimental to the health of especially susceptible groups of people such as the sick, the aged, and the very young. Air pollution forecast and warning systems coupled with legal authority have recently been instigated to prevent these episodic disasters.

Recent Research on the Effects of Air Pollution on Health

The U.S. Bureau of Mines Bulletin 537 has presented a bibliography of about 400 references on the effects of air pollution on health covering the period 1866 to 1954 [14]. The most recent comprehensive discussion of the medical and biologic effects of air pollution in Stern's handbook *Air Pollution* presents about 500 references including the year 1967 [15, 16]. For more than a hundred years medical evidence has been accumulated which demonstrates some harmful aspects of air pollution on health. Since the 1930s with the first major air pollution disaster in the Meuse Valley in Belgium, it has become clear that air pollution is a killer that can, under certain circumstances, take thousands of lives. These immediate or acute effects of air pollution are not questioned by anybody. It is the long-term or chronic effects of air pollution with their insidious results which cause all the controversy.

Before the medical profession can help, it must know what pollutants in the atmosphere are detrimental in what concentrations and combinations, and at what length of exposure. This job is exceedingly difficult, because with the high mobility of people, no sample group lives long enough in the same type of environment. Additionally, many pollutants by themselves in certain concentrations may be quite harmless. However, in combination they may react synergistically, i.e., the total harmful effect is much larger than the reaction of the chemicals acting independently. The cause-and-effect relationship between pollutants and disease or death is quite complex. The major research methods the medical sciences use are now briefly reviewed.

MEDICAL RESEARCH METHODS

Air pollution episodes. Acute air pollution episodes provide the most direct evidence of the effects of air pollution on health. If pollution and meteorological data have been obtained during an episode, comparison of the morbidity and mortality data with the air quality values can give valuable information for a pollution warning system. Safety precautions during a pollution alert would include indoor confinement and use of gauze masks.

Accidents. Industrial accidents such as those of Poza Rica and Cincinnati will happen again. Accidents from derailed trains carrying ammonia have occurred, and accidents involving nerve gas transports cannot be ruled out. Nuclear reactor accidents—so far 172 accidents in the United States alone since 1945—are certainly not rare. Medical research investigates carefully these "miniature" pollution disasters in order to develop counter measures.

Industrial research. Industrial hygienists have for a long time investigated the effects of harmful substances. In Table 3.3 industrial

Table 3.3 Industrial threshold limits and highest community pollution levels

Substance	Industrial threshold limits [17] ppm	Industrial threshold limits [17] mg/m^3	Highest community pollution levels	Place and date of observation
Carbon monoxide	50	55	360 ppm	London, England [18]
Lead	—	0.2	0.042 mg/m^3	Los Angeles, U.S., 1949-54 [19]
Nitrogen dioxide	5	9	1.3 ppm	Los Angeles, U.S., 1962 [20]
Ozone	0.1	0.2	0.9 ppm	Los Angeles, U.S., 1955 [20]
Sulfur dioxide	5	13	3.16 ppm	Chicago, U.S., 1937 [21]

Sources: American Conference of Governmental Industrial Hygienists, 1967 [17]; P. J. Lawther, et al., 1962 [18]; California State Dept. Public Health, 1955 [19]; Los Angeles County Air Pollution Control District Quarterly Contaminant Reports, 1955-1962 [20]; Stanford Research Institute Final Report, 1956 [21].

threshold limits for the most common pollutants are compared with the highest concentrations of these pollutants in community air. The industrial threshold limits are, however, in no way adequate for ambient air standards, because they are only based on healthy adult males and usually refer to an eight-hour exposure. Community threshold limits have to be based on those most affected, i.e., the sick, the elderly, and infants. It is interesting to note that the 1.23 ppm SO_2 level in the 1952 London air pollution disaster, which caused 4,000 excess deaths, was much lower than the values shown in Table 3.3. It should also be a matter of concern that community levels for carbon monoxide and ozone have already exceeded the industrial threshold limits.

Laboratory experiments. Animal experiments can clearly demonstrate a direct cause-effect relationship between certain pollutants and sickness or death. Figure 3.5 shows the various stages of development of skin tumor on mice, produced by benzo(a)pyrene, a polycyclic hydrocarbon, also found in the ambient air and in cigarette smoke. It usually takes very much larger quantities of these carcinogenic or cancer-producing substances than are normally present in the city air to produce these tumors. It is for this reason and because animal response to harmful substances is not necessarily the same as that of humans that the medical science is hesitant in speaking of a direct cause-effect relationship.

Epidemiological studies. These are carefully planned surveys studying the effects of air pollution on health for a whole community and over an extended period of time. The advantage over laboratory experiments is that people in their actual environment are studied. Usually the occurrence and distribution of morbidity and mortality due to pollution and weather factors is studied separately for age, sex, race, and occupation, etc. This technique can, for example, show the predominance of certain diseases in certain areas. But again it cannot establish a cause-effect relationship.

Clinical research. The other research approaches can be supplemented by clinical studies. Test persons have been subjected to certain concentrations of pollutants to measure, for example, the effect of pollutants on the muscular reaction time, the air capacity of the lungs, work performance on a bicycle ergometer, and the heart rate [22]. The effect of the Los Angeles type smog on athletic performance of high school cross-country track runners is shown in Figure 3.6. With increasing oxidant (smog) levels, the performance degrades rapidly [23].

54   ATMOSPHERIC POLLUTION

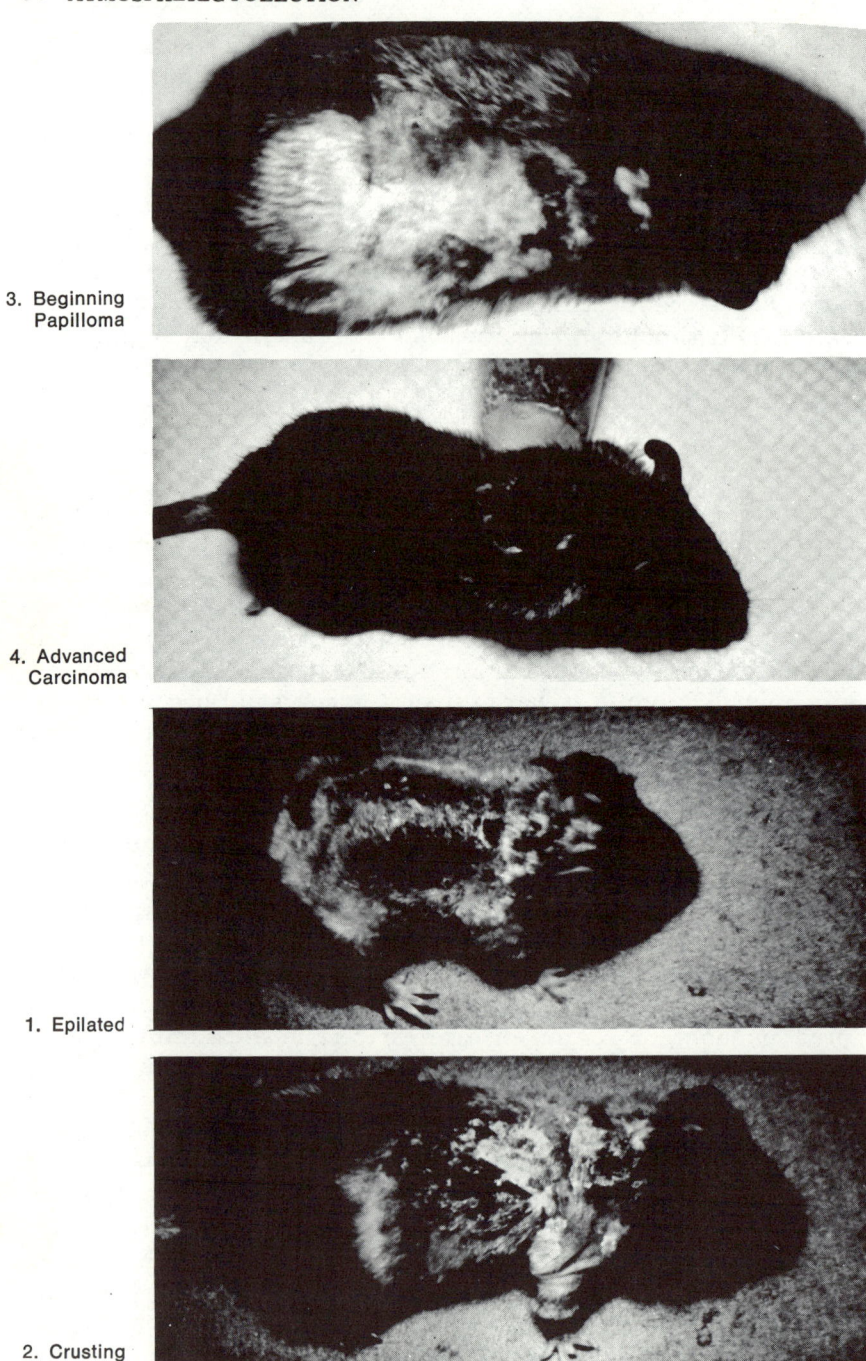

3. Beginning Papilloma

4. Advanced Carcinoma

1. Epilated

2. Crusting

Figure 3.5 Skin tumor development on mice (courtesy of Dr. Eula Bingham, Dept. of Environmental Health, Univ. of Cincinnati).

HEALTH ASPECTS OF AIR POLLUTION 55

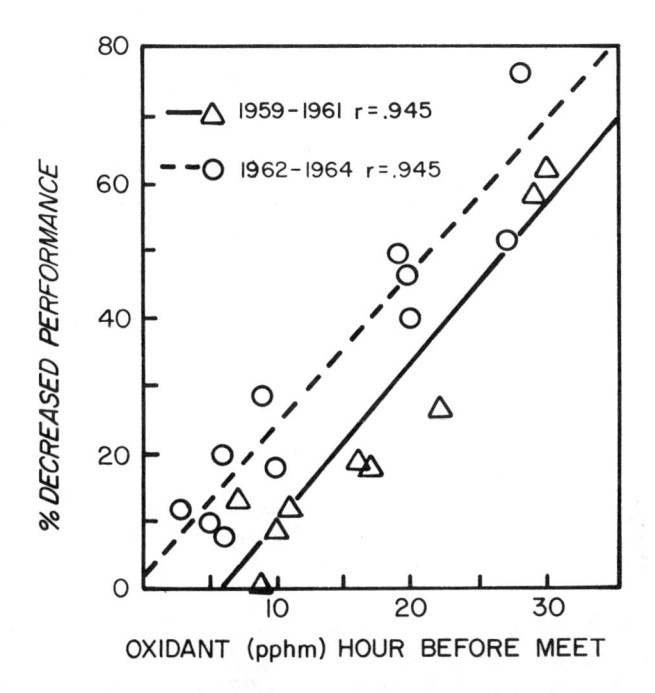

Figure 3.6 Oxidant level in the hour before the meet by percent of team members with decreased performance. (Source: W. S. Wayne, et al., 1967 [23])

MAJOR POLLUTANTS AFFECTING HEALTH

This section will discuss health effects of major air pollutants found in the air of cities. The threshold values extracted from a variety of investigations can be compared with the pollution levels in Table 3.3. This will give the reader a feeling for the severity of the problem.

A comparison of hospital admissions in Los Angeles for 223 consecutive days in 1961 with high pollution levels showed the following sequence of correlation coefficients: ozone (O_3) 0.27, sulphur dioxide (SO_2) 0.27, oxidants 0.25, nitrogen dioxide (NO_2) 0.19, particulate matter 0.16, and carbon monoxide (CO) 0.10 [24]. Correlation coefficients of 0.17 or greater were significant at the 1 percent level, and those of 0.13 or greater were significant at the 5 percent level. At least in the Los Angeles Basin the inorganic gases O_3, SO_2, oxidants, and NO_2 seem to constitute the greatest health hazard.

Ozone and photochemical oxidants. Ozone is a major component of photochemical oxidant, a substance which makes oxygen available for chemical reactions. O_3, NO_2, and hydrocarbons (HC) together with solar radiation form the Los Angeles type photochemical smog.

Odor detection of the pungent, colorless O_3 starts between 0.02 and 0.05 ppm, irritation of the nose and throat at 0.05 ppm, and dryness of the throat above 0.1 ppm [25]. Such levels of O_3 for periods longer than 30 minutes produce headaches. In Los Angeles oxidant levels of 0.1 ppm caused difficulty in breathing in patients suffering from emphysema [26]. The threshold limit for oxidants and O_3 should obviously be set below 0.1 ppm.

Sulfur dioxide. Taste detection of this colorless, pungent gas is at 0.3 ppm, the odor threshold lies at about 0.5 ppm [27]. Concentrations of 1.6 ppm over a few minutes produced broncho-constriction [28]. SO_2 is easily soluble in the nasal passages so that most of the time its irritant effects are restricted to the upper respiratory tract and the eyes. If, however, SO_2 becomes absorbed on small particles it will penetrate deeply into the lungs. This synergistic effect of a gas plus particulate matter was apparently responsible for the high excess death rate of the 1952 London episode.

Nitrogen dioxide. The odor threshold of this pungent gas, which produces a brownish haze, lies between 1 and 3 ppm; and nose and eye irritation has been associated with 13 ppm. At concentrations of 25 ppm volunteers complained of pulmonary discomfort after five minutes of exposure [29].

Carbon monoxide. This odorless and colorless gas is one of the most dangerous pollutants, because it is emitted in such large quantities at breathing level. CO poisoning can cause headache, dizziness, nausea, vomiting, difficulty in breathing, and unconsciousness, and finally death. The symptoms are produced by the fact that hemoglobin combines 240 times more readily with CO than with oxygen. CO thus prevents hemoglobin from transporting oxygen from the lungs to the tissues. The usual levels of 30 ppm CO found in city air bind about 5 percent of the hemoglobin, which in terms of oxygen supply of the blood is comparable to living at an altitude of 6,000 feet [15].

Lead. The average United States adult ingests about 0.3 milligram of lead (Pb) daily from food and water. The Pb taken in by breathing the city air is about 0.05 milligram per day [30]. Twenty to 50 percent of the inhaled Pb is usually retained in the body. It has been found that the higher the exposure to Pb the higher is the blood lead level. Often Pb is deposited in the marrow of the bone to become activated under physical and psychological stress. Thus any increase in the body burden of Pb is undesirable.

In the ambient atmosphere, which man breathes, the above pollutants do not occur isolated but in combination producing in

Table 3.4 Effect of auto exhausts on fertility and infant survival of mice

Components (ppm)					Total no. of litters	Total mice born	Average no. of mice per litter	No. of preweaning deaths
CO	HC	O_3	NO	NO_2				
20	6	0.2	0.4	0.3	135	995	7.4	656
50	18	0.5	1.3	1.2	106	729	6.9	607
60	20	0.6	1.5	1.4	63	431	6.8	357
100	36	1.0	2.0	1.9	54	370	6.8	309
Clean air					113	805	7.1	481

Source: F. G. Hueter, et al., 1966 [31].

many cases a synergistic effect, which means that the total effect is larger than the sum of the individual effects. Properly designed experiments will therefore analyze the synergistic effects. Table 3.4 shows the effect on mice of different concentrations of auto exhausts which have been irradiated by radiation to produce an air mixture comparable to photochemical smog. It is obvious that with increasing concentrations the mice fertility decreases [31].

MAJOR DISEASES RELATED TO AIR POLLUTION

Irritant pollutants, as we have seen, cause acute effects such as eye, throat, and nose irritation. Gases such as SO_2 usually dissolve in the sticky liquid called mucus which protects the air tracts (Figure 3.7). Gaseous and particulate irritants are intercepted by the cilia, which are hairlike cells that protect the air tracts. When the insult is too large, this protective mechanism breaks down and the pollutants penetrate deeply into the bronchioles and alveoli, the tiny air sacs deep in the lung (see Figure 3.7). If the insult continues year after year in large enough concentrations, then any one of the following major chronic diseases may develop.

Asthma. An asthmatic attack consists of the narrowing of the bronchioles, which is caused by a muscle spasm, an enlargment of the mucous membrane, and by abundant mucous secretions (see Figure 3.7). Asthma can be caused by allergens of natural origin such as pollen. Here we shall restrict ourselves to asthma caused by man-made pollutants. A widely publicized asthma attack related to SO_2 is that reported of Nashville, Tennessee, in 1961 [33]. The investigators concluded that the frequency of asthma attacks in adults, but not in children, was greater in more polluted areas. A similar asthma attack in New Orleans in October 1962 was blamed on a city dump and grain handling. This episode claimed 9 lives and hospitalized 300 persons.

58 ATMOSPHERIC POLLUTION

Bronchitis and emphysema. These two chronic respiratory diseases are discussed together, because they either occur side by side, or emphysema may be the follow-up of bronchitis. The diagnosis of bronchitis is justified, when the sufferer has a chronic cough with sputum lasting for at least four weeks and occurring for at least the previous three winters [34]. In emphysema the patient's air sacs or alveoli become over-extended and eventually break down (see Figure 3.7). Both bronchitis and emphysema induce shortness of breath in patients. In advanced stages they are unable to blow out a lighted match only a few inches away from their mouths.

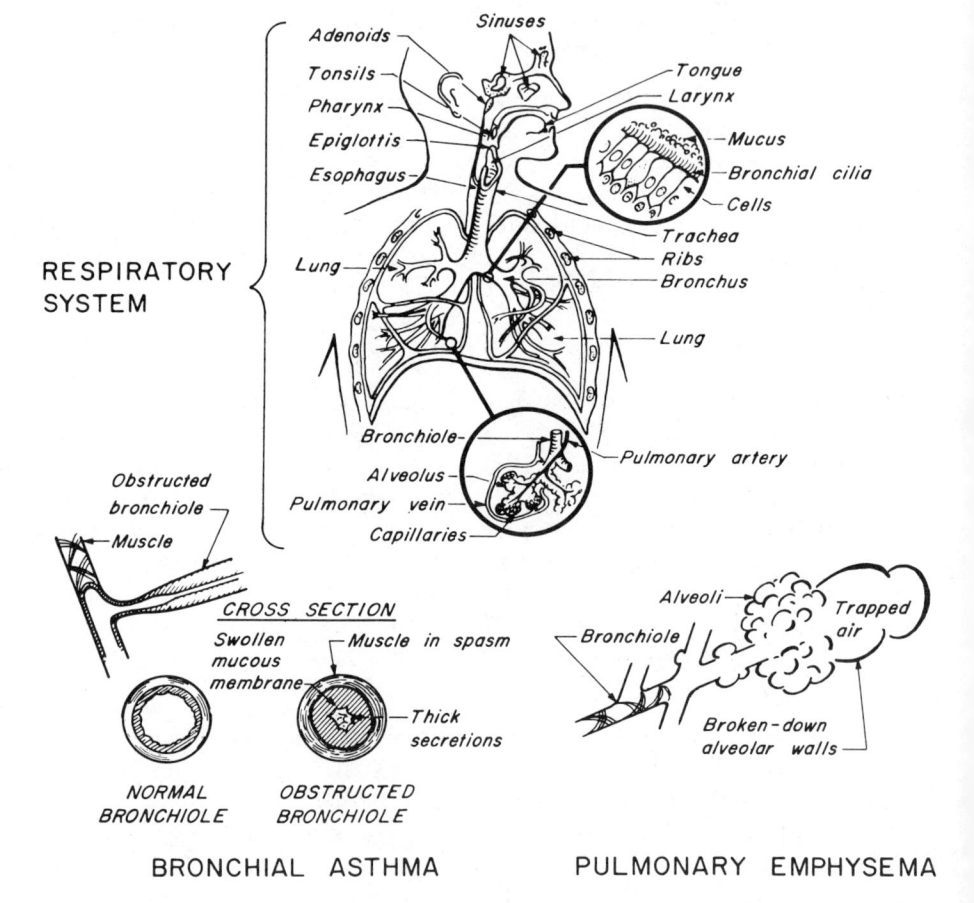

Figure 3.7 Respiratory system with bronchial asthma and pulmonary emphysema. (Source: National Tuberculosis and Respiratory Disease Association, 1969 [32])

A selection of respiratory diseases in California from 1950 to 1964 is shown in Figure 3.8. Whereas the bronchitis death rates remained more or less constant, it is quite obvious that the death rate of emphysema shows a manifold increase. Emphysema is indeed the fastest growing cause of death in the United States. One example, which stands for a large number of similar investigations, may suffice to demonstrate the deleterious effects of air pollutants on men with chronic bronchitis and emphysema. During the first and the third weeks of their hospital stay the patients breathed unfiltered air. During the second week all pollutants had been filtered from their breathing air. The results indicate a strong influence of the polluted air on the whole respiratory system. The heart had to work much harder to procure the oxygen supply. This additional strain on the cardiorespiratory system results in greater cardiovascular death rates.

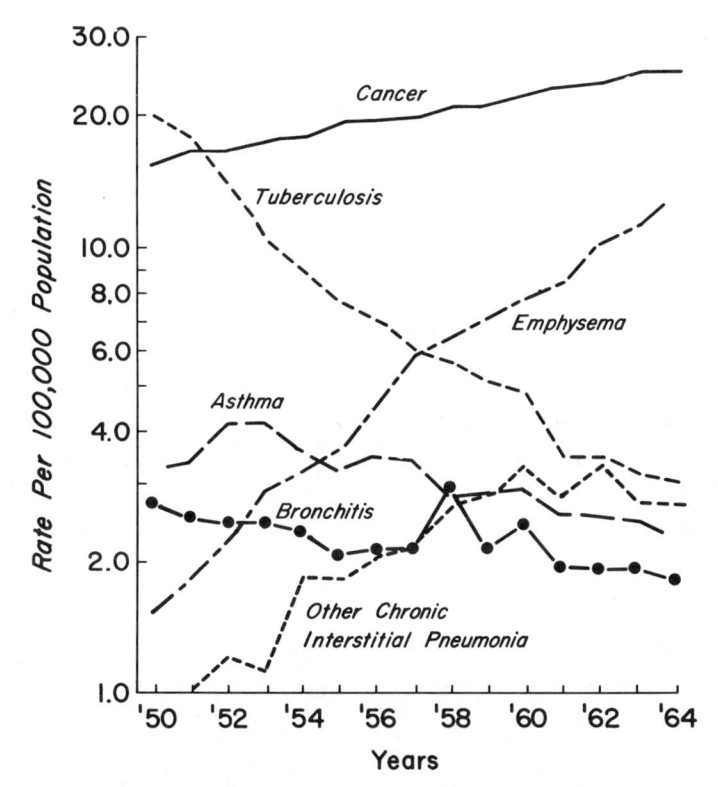

Figure 3.8 Death rates from selected respiratory diseases in California, 1950-1964. (Source: J. R. Goldsmith, 1968 [15])

Lung cancer. Cancer is produced by uncontrolled cell growth. Lung cancer is the abnormal and uncontrolled growth of cells which usually originates in the bronchial mucous membrane (see Figure 3.7). A large number of carcinogenic (cancer-producing) substances are known. The best-known is perhaps the 3,4 benzpyrene (Ba P), a polycyclic hydrocarbon (HC). HC are present in the air of cities and industrial areas. They result from incomplete combustion and are present in large quantities in automobile exhausts, cigarette smoke, and industrial combustion processes.

One fact is quite obvious, the death rate from lung cancer is rising as shown for a 14-year period in California (see Figure 3.8). This is also true for the United States as a whole, and it also holds for other countries, such as England and Wales. Table 3.5 shows that the death rate due to lung cancer has increased by a factor of 80 over a period of 56 years [34]. Also larger urban areas as compared with smaller urban and rural areas have a higher rate of lung cancer (Table 3.6).

Table 3.5 Lung cancer deaths in England and Wales, 1901-1956

Year	Death rate	Year	Death rate
1901	228	1931	1,880
1906	341	1936	3,432
1911	436	1941	5,084
1916	413	1946	8,110
1921	547	1951	13,247
1926	850	1956	18,186

Source: P. J. Lawther, 1959 [35].

Table 3.6 Death rates for lung cancer in England and Wales, 1950-1953

Population	Lung cancer Male	Lung cancer Female
Conurbation	126	121
Greater than 100,000	112	101
50,000 – 100,000	93	88
Less than 50,000	84	86
Rural	64	77

Source: R. E. Waller, 1959 [36].

In an attempt to show the direct relationship of HC and cancer, crude benzene extracts were injected into mice. Within a period of 9 to 14 months 2 to 10 percent of the mice developed tumors. The same research team also extracted HC from various city atmospheres and found that Birmingham, Alabama, produced the greatest tumor yield [37]. Incidentally Birmingham was also reported to have the highest death rate of human lung cancer among 163 cities [38].

It is now generally accepted that cigarette smoking is one likely causal factor in producing lung cancer. A "likely causal factor" is defined by Goldsmith as one "for which there is not substantial doubt as to causation" [15]. In regard to the effects of breathing the city atmosphere, there still seems to be a good deal of controversy as to the magnitude of the urban influence. The general consensus seems to be that the urban influence on lung cancer is real but that the evidence is not yet considered sufficient proof to relate it to air pollution [39, 40].

SUMMARY

The acute adverse effects of air pollution on health are fairly easy to demonstrate. It stands to reason also to expect chronic adverse effects of air pollution on health. The difficulty in obtaining a direct cause-effect relationship between air pollution and morbidity or mortality lies in the high mobility of people and their addiction to certain habits. Unless it can be shown which pollutants in the ambient air at what concentrations and combinations, and at what length of exposure, produce what kinds of adverse health effects, the medical profession is hesitant to speak of a direct cause-effect relationship between air pollutants and various diseases.

With the help of acute air pollution episodes, air pollution accidents, industrial hygiene research, laboratory experiments, epidemiological studies, and clinical research, one tries to shed further light on the direct cause-effect question. The major air pollutants such as ozone and photochemical oxidants, sulfur and nitrogen dioxides, carbon monoxide and lead, which are known to adversely affect health, are discussed. The pollutant concentrations and the symptoms they produce, are described.

Nobody denies that the atmosphere consists of irritant pollutants which can cause acute eye, throat, and nose irritation. It is furthermore recognized that air pollutants are a paramount factor in the development of the chronic diseases of asthma, bronchitis, emphysema, and lung cancer. The Surgeon General of the United States has called cigarette smoking a likely causal factor in producing the above diseases. Since the ambient air consists of substances similar to those found in cigarette smoke, it is logical to assume that air pollution is a contributing factor in the development of the above diseases.

REFERENCES CITED

[1] W. T. Russell, "The Relative Influence of Fog and Low Temperature on the Mortality from Respiratory Disease," *The Lancet*, 1128-1130, Nov., 1926.
[2] H. Heimann, "Effects of Air Pollution on Human Health," in World Health Organization, ed., *Air Pollution*, Monogr. Ser. No. 46, pp. 159-220, 1961.
[3] L. Greenburg, et al., "Report of an Air Pollution Incident in New York City, November 1953," *Publ. Health Rept.* 77 (7), 1962.
[4] A. E. Martin, "Mortality and Morbidity Statistics and Air Pollution," *Prcdgs. Roy. Med. Soc.* 57, 969-975, 1964.
[5] L. Greenburg, et al., "Air Pollution, Influenza, and Mortality in New York City," *Arch. Environ. Health* 15, 430-438, Oct., 1967.
[6] M. Glasser, et al., "Mortality and Morbidity During a Period of High Levels of Air Pollution," *Arch. Environ. Health* 15, 684-694, Dec., 1967.
[7] F. E. Speizer, "An Epidemiological Appraisal of the Effects of Ambient Air on Health: Particulates and Oxides of Sulphur, *JAPCA* 19(9), 647-655, Sept., 1969.
[8] R. E. Waller and B. T. Commins, "Episodes of High Pollution in London 1952-66," *Int. Clean Air Congr. Prcdgs.* 1, 228-231, London, Oct., 1966.
[9] J. R. Goldsmith, "Effects of Air Pollution on Human Health," in A. C. Stern, ed., *Air Pollution* 2nd ed., vol. 1, Academic Press, Inc., New York, 1968, pp. 554-563.
[10] H. H. Schrenk, et al., "Air Pollution in Donora, Pa. Epidemiology of the Unusual Smog Episode of October, 1948," *Publ. Health Bull.* (306), 1949.
[11] J. A. Scott, "The London Fog Disaster," *Prcdgs. 20th Ann. Clean Air Conf. Glasgow*, 25-27, 1953.
[12] J. McCarroll, "Measurements of Morbidity and Mortality Related to Air Pollution," *JAPCA* 17(4) 203-209, Apr., 1967.
[13] D. A. Lynn, et al., *The November-December 1962 Air Pollution Episode in the Eastern U.S.*, USDHEW, PHS Publ. No. 999-AP-7, Cincinnati, Ohio, Sept., 1964.
[14] S. J. Davenport and G. G. Morgis, *Air Pollution; A Bibliography*, U.S. Bureau of Mines Bulletin 537, 1954, 448pp.
[15] J. R. Goldsmith, "Effects of Air Pollution on Human Health," in A. C. Stern, ed., *Air Pollution*, 2nd ed., vol. 1, Academic Press, Inc., New York, 1968, 547-554, 563-615.
[16] H. E. Stokinger and D. L. Coffin, "Biologic Effects of Air Pollutants," in A. C. Stern, ed., *op. cit.*, 445-546.
[17] American Conf. Governm. Ind. Hygienists, *Threshold Limit Values for 1967*, Cincinnati, Ohio, 1967.
[18] P. J. Lawther, et al., "Carbon Monoxide in Town Air. An Interim Report," *Ann. Occup. Hyg.* 5, 241-248, 1962.
[19] California State Dept. Publ. Health, *Clean Air for California*, Berkeley, Cal., 1955.
[20] Los Angeles County Air Pollution Contr. District Quart. Contaminant Rpts., 1955-1962.
[21] Stanford Res. Inst., *Literature Review of Metropolitan Air Pollutant Concentrations—Preparation, Sampling and Essay of Synthetic Atmospheres*, Final Rpt., Menlo Park, Cal., 1956.
[22] G. J. Holland, et al., "Air Pollution Simulation and Human Performance, *Am. J. Publ. Health* 58(9), 1684-91, Sept., 1968.

[23] W. S. Wayne, et al., "Oxidant Air Pollution and Athletic Performance," *J. Am. Med. Assoc.* 199(12) 151-154, Mar., 1967.
[24] T. D. Sterling, et al., "Measuring the Effect of Air Pollution on Urban Morbidity," *Arch. Environ. Health* 18, 485-494, Apr., 1969.
[25] A. Henschler, et al., "Olfactory Threshold of Some Important Gases and Manifestations in Man by Low Concentrations," *Arch. Gewerbepathol. und Gewerbehyg.* 17, 547-570, 1960.
[26] J. E. Remmers and O. J. Balchum, "Effects of Los Angeles Urban Pollution upon Respiratory Function of Emphysematous Patients," presented at AMA Air Poll. Med. Res. Conf., Los Angeles, 1966.
[27] F. I. Dubrovskaya, "Hygiene Evaluation of Pollution of Atmospheric Air of a Large City with SO_2," in V. A. Ryazanov, ed., *Limits of Allowable Concentrations of Atmospheric Pollutants* (transl. by B. S. Levine), U.S. Dept. Commerce, Washington, D.C., 1957.
[28] Y. Tomono, "Effects of SO_2 on Human Pulmonary Functions," *Japan J. Ind. Health* 3, 77-85, 1961.
[29] F. H. Meyers and C. H. Hine, "Some Experiences of NO_2 in Animals and Man," presented at 5th Air Poll. Med. Res. Conf., Los Angeles, 1961.
[30] T. Bersin, "Schädigungen von Enzymsystemen durch toxische Verbindungen der Autoabgase," *Vitalstoffe und Zivilisationskrankheiten* 11(55), 207-210, 1966.
[31] F. G. Hueter, et al., "Biological Effects of Atmospheres Contaminated by Auto Exhaust," *Arch. Environ. Health* 12, 553-560, May, 1966.
[32] *Air Pollution Primer*, Nat. Tub. Resp. Dis. Assoc., New York, 1969, pp. 55-76.
[33] L. D. Zeidberg, et al., "Nashville Air Pollution Study, I. SO_2 and Bronchial Asthma. A Preliminary Report," *Amer, Rev. Resp. Dis.* 84, 489-503, 1961.
[34] J. Pemberton and C. Goldberg, "Air Pollution and Bronchitis," *Brit. Med. J.* 4887, 567-70, 1954.
[35] P. J. Lawther, "Cancerogen in der städischen Luft," *Zeitschrift für Bakteriologie*, 1, Abt. 176, Originale, 187-193, 1959.
[36] R. E. Waller, *Air Pollution as an Etiological Factor in Lung Cancer*, Publ. of Acta Union Int. contre le Cancer, 15, 437, 1959.
[37] W. C. Hueper, et al., "Carcinogenic Bioassays on Air Pollutants," *Arch. Pathol.* 74, 89-116, 1962.
[38] N. E. Manos, *Comparative Mortality Among Metropolitan Areas in the US, 1949-1951*, PHS. Publ. No. 562, 1957.
[39] E. L. Wynder and E. C. Hamond, "A study of Air Pollution Carcinogenesis," *Cancer* 15, 79-92, 1967.
[40] P. Kotin and H. L. Falk, "Atmospheric Factors in Pathogenesis of Lung Cancer," in A. Haddow, ed., *Advances in Cancer Research*, vol. 7, Academic Press, Inc., New York, 1963, pp. 475-514.

CHAPTER 4

ECONOMIC ASPECTS OF AIR POLLUTION

There is no longer any doubt that air pollution places a grave burden upon the national economy as well as upon the economy of individual families. Air pollution soils and erodes building surfaces, corrodes metals, weakens textiles, deteriorates works of art; more importantly, it damages vegetation and crops, kills animals, and drastically interferes with the well-being of people. The existence of all these adverse effects should provide incentive enough for reducing the pollution of the atmosphere. In a materialistic world such as ours, however, altruistic motivations alone are apparently not sufficient to combat air pollution. Stronger incentives, such as putting a price tag on the natural resource air, seem to be necessary.

Many analyses of the economic aspects of air pollution discuss only the cost of damages caused by pollution and the amounts spent by pollutors for control equipment. A true assessment of the economic impact of air pollution, however, can be obtained only if the cost of damages and control efforts are related to the value of the benefits to be gained by control.

COST OF AIR POLLUTION DAMAGES

Property. A recent investigation revealed that in the St. Louis metropolitan area, house values decreased by $245 for every increase of 0.5 milligram of SO_3 per 100 square centimeters per day [1]. Houses sold in St. Louis for $10,000 in 1957 were worth only $8,977 in 1964, a total loss of $1,023 over the seven-year period. Based on the number of damaged dwelling units in metropolitan St. Louis, a total estimated loss of $765,000 occurred.

In another study the costs of air pollution effects between heavily polluted Steubenville, Ohio (annual mean suspended particulate concentration of 383 $\mu g/m^3$) and relatively clean Uniontown, Pennsylvania (annual mean suspended particulate concentration of 115 $\mu g/m^3$) were compared [2]. It was shown that the per capita annual costs for outside and inside maintenance of houses, laundry and dry cleaning, and hair and facial care were $84 higher in Steubenville than in Uniontown.

Other examples of direct property damage through air pollution are reported by Ridker [2]. In 1962 pungent gases from a new metal fabricating plant started to drift into a residential neighborhood in St. Louis. Comparison of property value indices in the polluted area with those in unpolluted areas showed that houses sold after 1962 in the affected areas suffered a loss of about $1,000. In Syracuse, New York, a power plant released soot over only a 20-minute period. Shortly after the incident citizens of the affected neighborhood were interviewed in order to assess additional cleaning costs and insurance claims. Estimates indicate that during this short time period the emission of suspended particulates could only have been 225 pounds above that which would normally be released. Apparently this small amount of pollutants was sufficient to cause $38,000 in damages.

Traffic. Pollution episodes with dense fogs reduce visibility, slow down traffic, and cause costly accidents. People reach their working places with delays and have to work more hours to make up for the loss. Long traffic delays may be deleterious to perishable foods, particularly fresh fruit and vegetables.

On request the British Transport Commission, London, released the information that one fog day causes them additional costs of about $10,000 due to loss of income or pay for extra personnel [4]. One airline alone, the British European Airways, suffered a loss of about $800,000 between November 1958 and February 1959 due to flight cancellations on account of thick fog. It is well-known that increased air pollution results in decreased visibility. Thus, to a large extent air pollution is responsible for the traffic industry's extra costs.

Agriculture. In 1951 a leaf spot disease hit the tobacco growers in Connecticut [5]. It was later discovered that this disease was due to ozone, a major ingredient of photochemical smog, produced through the reaction of auto exhausts with solar radiation. In 1957 the Connecticut farmers lost an estimated $1 million worth of cigar wrapper leaf. In 1959 a single smog weekend resulted in a loss of $6 million.

Not only tobacco, but other valuable plants such as cotton, beans, lettuce, tomatoes, grapes, citrus, and several pine species are very susceptible to smog damage. The total damage each year to crops in the United States is estimated to amount to $500 million [6]. One-fourth of this damage bill, about $125 million, has to be paid by smog-ridden California alone.

Health. For obvious reasons it is very difficult to assess health costs related to air pollution. In order to assess the total cost to the community from diseases such as chronic bronchitis and emphysema leading to morbidity and mortality, Bates has suggested the acquisition of the following information [7]:

1. An assessment of the loss of productive working hours in industry in workers with chronic bronchitis compared to workers without this disease.
2. An assessment of the loss of working potential as a result of the crippling power of emphysema and bronchitis.
3. An assessment of the welfare compensations caused by these diseases.

A study based on these criteria was conducted in the United Kingdom. It was shown that chronic bronchitis caused the loss of 26.6 million days of work among the insured population in 1951. At average earnings of $20 per day, this would amount to a loss of $532 million. If 70 percent of this loss can be attributed to cigarette smoking, 20 percent to air pollution, and 10 percent to miscellaneous other causes, then the loss due to air pollution would still amount to a staggering $106 million. It must be realized that this very conservative estimate relates to British pay rates in the 1950s and considers only bronchitis and emphysema. Taking all diseases caused or aggravated by air pollution into consideration and allowing for the four times greater number of people with higher pay scales, the health costs attributable to air pollution would probably approach the $1 billion mark in the United States.

Total estimated economic losses. Since it is so difficult to assess health costs of diseases related to air pollution, cost estimates usually deal only with tangible losses. As early as 1913 Pittsburgh made a cost assessment which revealed that the per capita cost per year due to air pollution amounted to $20. This figure did not include health costs inflicted by air pollution.

The 1952 London air pollution disaster shocked the British nation into forming an investigating committee led by Lord Beaver [8]. The Beaver Report of 1954, whose recommendations led to the British Clean Air Act in 1956, estimated a total economic loss from air pollution of more than $1 billion per year (Table 4.1). This amounts to about $35 per capita per year in urban and industrial areas, and to about $15 per capita annually if related to the total

British population. And yet these startling figures include neither health costs nor the estimated loss of 50 million working days through illness and premature crippling.

Table 4.1 Costs of air pollution in millions of dollars per year in Great Britain

Cost item	Millions of $ per year
Laundry	125
Painting and decorating	150
Cleaning and depreciation of buildings other than houses	100
Corrosion of metals	125
Damage to textiles and other goods	262.5
Total direct costs	762.5
Loss in efficiency	500
Total economic loss	1,262.5
Cost per capita in industrial areas	$ 35 per year
Cost per capita of total population	$ 15 per year

Source: H. Beaver, 1958 [8].

The estimates for total economic losses due to air pollution in the United States range from $2 to $12 billion per year depending on what is included in the estimates [9]. Health considerations are not included in the above estimates, although they indeed constitute the greatest economic losses. Table 4.2 summarizes the air pollution costs in millions of dollars by states and per year [10].

THE COST OF CLEANER AIR

Two contrasting opinions exist when it comes to the question, Who is to pay for cleaner air? (1) One group argues that if the total air pollution control costs are not lower than the total benefits resulting from controlling air pollution, then investing in control devices could not be considered sound economics either from the company's or the public's standpoint. (2) The second group states bluntly that the cost of pollution should be imposed on the pollutor. It is argued that if the pollutors were forced to internalize their external social pollution costs, they would develop methods of controlling the pollution. The following section of examples from a variety of business branches will give further insight into the cost aspects of cleaner air.

Chemical companies. By means of its control devices the Hood Sponge Rubber Company in Chicago, Illinois, recovers 1483 gallons of acetone, freon TF perchlorethylene, and other chemicals monthly [11]. The management finds it gratifying that the pollution control system saves the company $39,000 annually.

68 ATMOSPHERIC POLLUTION

Table 4.2 Air pollution costs in thousands of dollars per year in the United States

State	1964 Population	Cost of pollution	Being spent now	Cost of control
Alabama	3,414	$ 221,910	$ 8,535	$ 53,258
Alaska	256	16,640	640	3,994
Arizona	1,598	103,870	3,995	24,929
California	17,947	1,166,555	44,868	279,973
Colorado	1,978	128,570	4,945	30,857
Connecticut	2,683	174,395	6,708	41,855
Delaware	485	31,525	1,213	7,566
District of Columbia	785	51,025	1,963	12,246
Florida	5,751	373,815	14,378	89,716
Georgia	4,180	271,700	10,450	65,208
Hawaii	715	46,475	1,788	11,154
Idaho	700	45,500	1,750	10,920
Illinois	10,457	679,705	26,143	163,129
Indiana	4,825	313,625	12,063	75,270
Iowa	2,806	182,390	7,015	43,774
Kansas	2,244	145,860	5,610	35,006
Kentucky	3,073	199,745	7,683	47,939
Louisiana	3,394	220,610	8,485	52,946
Maine	1,003	65,195	2,508	15,647
Maryland	3,300	214,500	8,250	51,480
Massachusetts	5,301	344,565	13,253	82,696
Michigan	8,133	528,645	20,333	126,875
Minnesota	3,557	231,205	8,893	55,489
Mississippi	2,261	146,965	5,653	35,272
Missouri	4,402	286,130	11,005	68,671
Montana	711	46,215	1,778	11,092
Nebraska	1,483	96,395	3,708	23,135
Nevada	388	25,220	970	6,053
New Hampshire	643	41,795	1,608	10,031
New Jersey	6,430	417,950	16,075	100,308
New Mexico	1,029	66,885	2,573	16,052
New York	17,676	1,148,940	44,190	275,746
North Carolina	4,771	310,115	11,928	74,428
North Dakota	644	41,860	1,610	10,046
Ohio	10,369	673,985	25,923	161,756

(continued)

Table 4.2 (continued)

State	1964 population	Cost of pollution	Being spent now	Cost of control
Oklahoma	2,477	$ 161,005	$ 6,193	$ 38,641
Oregon	1,887	122,655	4,718	29,437
Pennsylvania	11,504	747,760	28,760	179,462
Rhode Island	878	57,070	2,195	13,697
South Carolina	2,458	159,770	6,145	38,345
South Dakota	709	46,085	1,773	11,060
Tennessee	3,668	238,420	9,170	57,221
Texas	10,313	670,345	25,783	160,883
Utah	985	64,025	2,463	15,366
Vermont	393	25,545	983	6,131
Virginia	4,237	275,405	10,593	66,097
Washington	3,030	196,950	7,575	47,268
West Virginia	1,748	113,620	4,370	27,269
Wisconsin	4,136	268,840	10,340	64,522
Wyoming	351	22,815	878	5,476
Totals	188,166	$ 12,230,790	$ 470,429	$ 2,935,392

Permission granted to reprint from *Materials Protection*, Vol. 6, No. 5, 47-52, May, 1967, published by the National Association of Corrosion Engineers.

W. R. Grace & Co. in Cincinnati, Ohio, installed a wet scrubber system consisting of 48,000 perforated plastic balls which absorb fine dust and ammonia vapor [12]. During the first nine months of installation more than $60,000 were saved in recovered products. Since the control device cost $102,000, it will pay for itself in about 15 months. Prior to control the plant emitted 130 pounds per hour of fine dust into the atmosphere. The control device reduced the emission to 30 pounds per hour. Apart from the profitable recovery of the material, a significant contribution to cleaning the ambient air was made.

Another chemical company installed a dust collection system which recovers annually about 4,000 tons of fly ash. Since this material contains about 10 percent carbon it can be reused as fuel. Another plant of this same company installed a $5,000 control device to recover the foul-smelling benzothiazole. The company now recovers about $50,000 worth of this material a year.

Utility plants. A large utility plant used to collect fly ash in precipitators and to dispose of the more than 1,000 tons daily at a disposal cost of $1.75 per ton. Now, a cindering plant pelletizes the ash, selling the new product at $4.50 per ton. Prior to control the plant paid $1,750 per day disposal costs, whereas now the sold material brings $4,500 daily. This amounts to a net profit of $6,250 per day, or $2,281,250 per year.

When a New York City ordinance restricted the sulfur content of fuel burnt in power plants, the additional annual costs were estimated to amount to $50 million [10]. The companies quickly passed their costs on to the customer, which meant for each New York family an additional $5 per year. Yet the utility companies boasted that they had spent $118 million for air pollution control. In 1967 when Consolidated Edison in New York switched from coal with a 1.6 percent sulfur content to coal with a 1 percent sulfur content, the coal price increased by $0.04 to $0.37 per mm Btu.

Linsky has presented a graph of cost estimates for electricity consumption in a midwestern city (Figure 4.1) [13]. Total customer costs, distribution, fuel, operation, maintenance, air pollution control, and other costs can now be viewed in proper perspective. It can be seen that the extra cost due to air pollution control at $0.06 per month (only the 156th part of the total cost of $9.40) is indeed quite minimal.

Smelters. The production cost of utilizing the by-product sulfuric acid (H_2SO_4) from nonferrous smelters is $4.25 per ton whereas its production from elemental sulfur would cost $10 to 20 per ton [14]. Smelters are thus a competitive source of H_2SO_4. The 26 smelters west of the Mississippi could produce about 7.3 million tons of H_2SO_4. This would almost be as much as the 7.4 million tons of H_2SO_4 that were actually consumed in 1965 in the 22 states west of the river. In reality only 2.8 million tons of H_2SO_4 from the smelters were absorbed by the market. High transportation costs were suspected for the small usage. It was suggested that if the smelters could sell their H_2SO_4 for $4 per ton FOB the smelter, all of it would find a market, and at the same time the current sulfur emissions would be reduced by 60 to 65 percent.

Petroleum refineries and motor vehicles. In 1966 the oil refineries released into the air some 2.1 million tons of sulfur oxides [15]. This amounted to about 7 percent of the total national sulfur oxide emissions. It has been estimated that in 1974 the annual cost of maximum control of particulate matter and 90 percent control of sulfur oxides would amount to $18 million for 85 metropolitan areas. This amount would only be 0.14 percent of the projected value of petroleum in that year.

Reduction of pollution emission from motor fuels in the United States can be achieved by lowering the volatility and by eliminating

ECONOMIC ASPECTS OF AIR POLLUTION 71

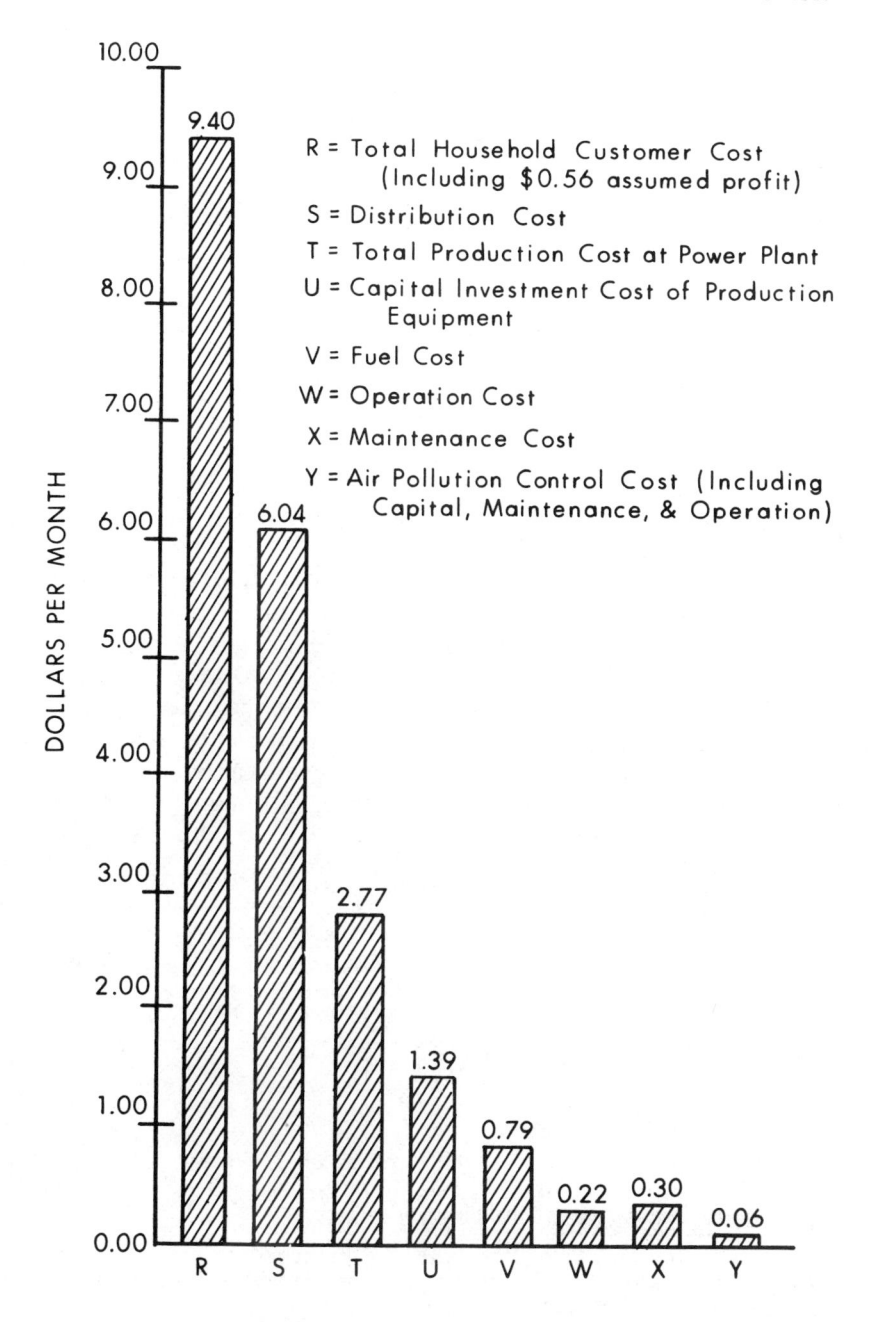

R = Total Household Customer Cost (Including $0.56 assumed profit)
S = Distribution Cost
T = Total Production Cost at Power Plant
U = Capital Investment Cost of Production Equipment
V = Fuel Cost
W = Operation Cost
X = Maintenance Cost
Y = Air Pollution Control Cost (Including Capital, Maintenance, & Operation)

R: 9.40
S: 6.04
T: 2.77
U: 1.39
V: 0.79
W: 0.22
X: 0.30
Y: 0.06

Figure 4.1 Average household customer s monthly price of electricity at 375 kw-hr per month. (Source: B. Linsky, 1966 [13])

72   ATMOSPHERIC POLLUTION

the olefins, the unsaturated hydrocarbons, which react with nitrogen oxides and solar radiation to form the photochemical smog. United States refineries could turn out such products if they invested $1.83 billion [16]. This figure sounds less frightening if one relates it to a unit price per gallon of gasoline, which is exactly what the oil companies do, when they hand on to the consumer all pollution control or other costs. Thus a decrease in volatility of 5 Rvp (Reed vapor pressure), which is about one-half the present level, would only add 1.6 cents per gallon to the cost of gasoline.

In a statement before the Subcommittee on Air and Water Pollution of the Senate Committee on Public Works in Washington, D.C., on March 18, 1970, B. Linsky stated: "I have become fed up with the publicly scary figures that result when the added air pollution control capital costs and annualized costs of something I buy is given an emotional aura by multiplying it by millions of buyers and multiplying that hobgoblin figure again by several years of purchases and usage" [17].

Table 4.3 presents such a tabulation [15]. The capital costs are the automobile industry's estimates of the costs that have been or will be passed on to new car buyers. It is important to note that the capital costs handed on to the consumer do not reflect the actual production costs which are kept secret by the automobile manufacturers. The annualized costs are those related to the operation and maintenance of the pollution control device. Table 4.3 shows that the costs of all control devices and their maintenance for all cars in the nation over a seven-year period would cost almost $3 billion.

Table 4.3 Estimated cost of vehicle emission control systems 1968-1974

	Assumed cost per car		Estimated national cost (millions of dollars)			
			Capital cost		Annualized	
	Capital cost	Annualized	Annual	Cumulative	Annual	Cumulative
1968	$18.00	$1.80	$180	$	$18	
1969	18.00	1.80	180	360	18	$ 36
1970	36.00	3.60	360	720	36	72
1971	48.00	5.80	480	1,200	58	130
1972	48.00	5.80	480	1,680	58	188
1973	48.00	5.80	480	2,160	58	246
1974	48.00	5.80	480	2,640	58	304

Source: First Report, Secretary HEW, 1969 [15].

What meaning does this scare figure have? The automobile industry does not pay a single cent of this amount. The individual car owner pays the bill for air pollution control, exactly $48.00 for a new car in

1971 and $5.80 for maintenance of the control device. In compliance with the motor vehicle emission standards of the Clean Air Amendments of December 1970, from 1971 onwards each owner of a new car will contribute $53.80 to auto emission control. The automobile industry will make the profit by manufacturing and installing the control devices.

It has been shown above that the cost of clean air is shared by many. It is also now clear that pollutors can derive major benefits from air pollution control which is manifested in the production of control devices, the sale or reuse of reclaimed effluents, in reduced damage to materials and equipment, in an improved working environment, and in the avoidance of fines imposed by legislative action. The major cost sharing in air pollution control, however, is that between government and industry as discussed below.

GOVERNMENTAL COST SHARING IN AIR POLLUTION CONTROL

According to Wilson and Minnotte "the effect of the government cost sharing is to provide a kind of insurance policy which guarantees that industry will not have to pay more than 41 percent of the cost of air pollution control" [18]. The governmental "insurance policy" includes federal corporate income tax, depreciation allowances, investment credits (rescinded for property contracted after April 18, 1969), Small Business Administration loans, Economic Development Administration aid, and state tax laws. The following discussion will deal with these provisions in detail.

Federal corporate income tax. For all income above $25,000 the Federal government obtains 48 percent of the corporation's net revenue. The sharing in net revenue is well known. The sharing of the government in the corporation's expenses, however, is usually not known. An example can best demonstrate the government's cost sharing. A corporation with a net income (revenue less expenses) of $2 million will pay 48 percent income tax (22 percent normal tax and 26 percent surtax), or $960,000. If pollution-control equipment worth $200,000 is purchased, the net revenue will become $1.8 million, because the control equipment is tax deductible. Instead of $960,000, now only $864,000 is paid in taxes. The Federal government, i.e., the general public, is paying 48 percent, or $96,000, of industry's total bill of $200,000 for air pollution control.

Depreciation allowances. The Internal Revenue Service makes depreciation allowances for capital equipment expenditures such as air pollution control devices. Instead of an equal depreciation over the total lifetime of the equipment which is usually 15 years, for tax purposes corporations make use of an accelerated depreciation method. For example, the depreciation tax savings from a pollution control system worth $4.08 million were calculated by Wilson and Minnotte to amount to $244,800 after the first year and only to $16,320 after the 15th year [18]. The difference between an equal and an accelerated depreciation acts as an interest-free loan to the corporation.

Investment tax credits. A 7 percent credit is given by the Federal government on capital investments which also include air pollution control devices. Taking the same example as above, the investment credit of 7 percent on the pollution control equipment cost of $4.08 million would result in tax savings of $285,600. Since this credit is not given in equal increments over the whole 15 years of the life of the equipment, but rather in the year in which the device is purchased, the investment credit, like the accelerated depreciation allowances, acts again as an interest-free loan. The Tax Reform Act of 1969 repealed the investment tax credit. It now applies only to property contracted for prior to April 18, 1969.

Small Business Administration loans. The idea is to avoid imposing hardships or possible insolvency upon smaller firms when they have to purchase control equipment in compliance with the Clean Air Act. About 90 percent of the pollution-producing industry is eligible for Small Business Administration loans, which guarantee access to low interest money.

Economic Development Administration. This agency offers technical and financial assistance of up to 100 percent of the total costs to any corporation, regardless its size, in order to control air pollution effectively. The only prerequisite is that the pollution source to be controlled has to be in an economically depressed area. But then almost one-third of the land area in the United States is so defined.

State taxes. States impose an income tax on industry's profits ranging from 2 percent to 11 percent. Like the Federal tax laws the state tax laws exempt control equipment from taxation. Since state taxes are deductible from income already taxed federally, state taxes result in actual average savings of 2.75 percent for the corporation.

The total share of industry's cost of air pollution control borne by the Federal government has been calculated by Wilson and Minnotte to amount to 59 percent [18]. In their conservative estimates they implied that the corporations would not pass on the cost of air pollution control to the general public. In reality, of course, these additional costs to industry are passed on to the consumer so that the public actually pays in a number of ways:

indirectly by cost sharing the taxes, directly by paying higher prices, and of course, in terms of the arbitrary effects of pollution itself.

COST-BENEFIT ANALYSIS OF AIR POLLUTION CONTROL

Air pollution control is a classic example of an external diseconomy. Whenever market forces are insufficient to make an individual bear the private and social costs resulting from his actions, such as polluting the atmosphere, external diseconomies arise. The external diseconomies can be eliminated, if the social costs can be internalized, i.e., if industry which is producing the pollution is also required to pay for its elimination or removal.

The optimum economic level of air pollution control, which does not necessarily coincide with the optimum health level of air pollution control, can be assessed by a cost-benefit analysis. Cost-benefit models have been developed for the St. Louis airshed [19], for the Delaware Valley transportation sources [20], and as a training exercise for air pollution control officers [21]. The cost-benefit analysis presented here was developed by Wilson and Minnotte [22]. They give an integrated picture of the cost for the benefits from an optimum level of particulate pollution in the Washington, D.C., area.

Cost-benefit model. The purpose of this cost-effectiveness analysis is to determine the maximum net benefit from air pollution control. The difference between the costs of air pollution control and the benefits from a reduction in air pollution constitutes the net benefit to a community.

The total net benefit (P_T) is related to the total benefit (B_T) and the total cost of control (CC_T) as follows:

$$(1) \qquad P_T = B_T - CC_T$$

Since there are many pollution emitters in an urban area, the total cost of control (CC_T) is simply the summation of the individual costs of control (ΣCC_j) from the n emitters. Thus,

$$(2) \qquad CC_T = \sum_{j=1}^{n} CC_j$$

Since each pollution emitter has some effect on each receptor area, the total benefit (B_T) is the summation of the benefit (B_{ji}) received in the i-th of m receptor areas due to the control of the j-th of the n emitters. Thus,

76 ATMOSPHERIC POLLUTION

(3) $$B_T = \sum_{i=1}^{m}\sum_{j=1}^{n} B_{ji}$$

Combining equations (1) to (3) yields the net total benefits (P_T):

(4) $$P_T = \sum_{j=1}^{n}\left(\sum_{i=1}^{m} B_{ji} - CC_j\right)$$

The equation was written in this form in order to emphasize that part of the equation in parenthesis which allows calculation of the net benefit from the control of each emitter.

In order to apply the cost-benefit model to any urban area, two sets of input data have to be obtained: the costs of control, and the value of the benefits from air pollution control.

Costs of air pollution control. The model is now applied to the Washington, D.C., area which includes the District of Columbia, a few counties in Maryland and Virginia, and the cities of Alexandria, Falls Church, and Fairfax. In interpreting the results it is important to remember that for lack of other data this model considers only suspended particulates, average annual and not short-period pollution levels, stationary point but no area sources, and finally, residential soiling as the only index of damages.

First an emission inventory was conducted for all sources of suspended particulates for which control costs were to be determined. Then alternatives for an emission control were considered including low-ash coal and low-sulfur oil, natural gas, mechanical controls, scrubbers, and electrostatic precipitators. The procedure was then to rule out nondominant control steps, which meant that only the least expensive control mechanisms for reducing particulate emissions were considered. Regarding the overall costs due consideration was given to purchase, installation, maintenance, and operating costs for mechanical control devices, fuel costs, and boiler conversion costs for switching to low-ash and low-sulfur fuels. For each of the 131 pollution emitters in the area the control costs were determined. For each emitter a table was prepared for the least-cost control alternatives versus particulate emission levels.

This procedure is illustrated for a power plant in Table 4.4 and for a municipal incinerator in Table 4.5. The power plant in Table 4.4 which presently emits 230 tons of suspended particulates per year burns a 2 percent sulfur and a 8.7 percent ash coal. It is equipped with an electrostatic precipitator of 98 percent operating efficiency. At this base level the control cost is considered as zero. Four emission controls are now considered. The first control step would upgrade the efficiency of the existing precipitator to 99.5 percent. This would reduce the particulate emissions to 58 tons per year at an annual cost of $33,900. The next three control steps only involve the replacement of the 8.7 percent ash coal by a 8.1 percent, 7.5 percent, and 6.9 percent ash fuel, respectively. It is obvious that small pollution reductions now results in soaring control costs.

Table 4.4 Cost-benefit analysis for a power plant

	Particulate emissions tons/year	Control cost $/year	Benefits $/year	Net Benefit $/year
	230	0	0	0
Four	58	33,900	196,000	162,000
control	53	102,300	200,600	98,300
steps	49	204,800	205,200	400
	45	482,500	209,700	−272,800

Source: R. D. Wilson and D. W. Minnotte, 1969 [22].

In the case of the municipal incinerator in Table 4.5 five control steps are considered. At the zero control cost level the incinerator has no pollution control devices. At a low annualized cost of $19,800 a baffled spray chamber can reduce particulate emission from 390 to 195 tons per year. The next control measures, adding first a cyclone collector, then an electrostatic precipitator, followed by a wet scrubber and a higher efficiency scrubber, may reduce particulate emissions to levels as low as 16 tons per year at relatively small annualized control costs. We have now to determine at which particulate emission levels the greatest net benefits are achieved.

Table 4.5 Cost-benefit analysis for a municipal incinerator

	Particulate emissions tons/year	Control cost $/year	Benefits $/year	Net benefit $/year
	390	0	0	0
	195	19,800	743,000	723,200
Five	86	27,800	1,159,000	1,131,200
control	78	43,600	1,188,800	1,145,200
steps	20	60,500	1,411,700	1,351,200
	16	67,200	1,427,500	1,360,300

Source: R. D. Wilson and D. W. Minnotte, 1969 [22].

Benefits of air pollution control. The benefits from air pollution control are obtained from the assessment of damages caused by air pollution in the absence of control. For the Washington, D.C., area only residential soiling costs were available; these were obtained from

78 ATMOSPHERIC POLLUTION

relating cleaning and maintenance operations with levels of suspended particulates. Figure 4.2 shows for the District of Columbia, Alexandria, and Fairfax City the straight line relationship between

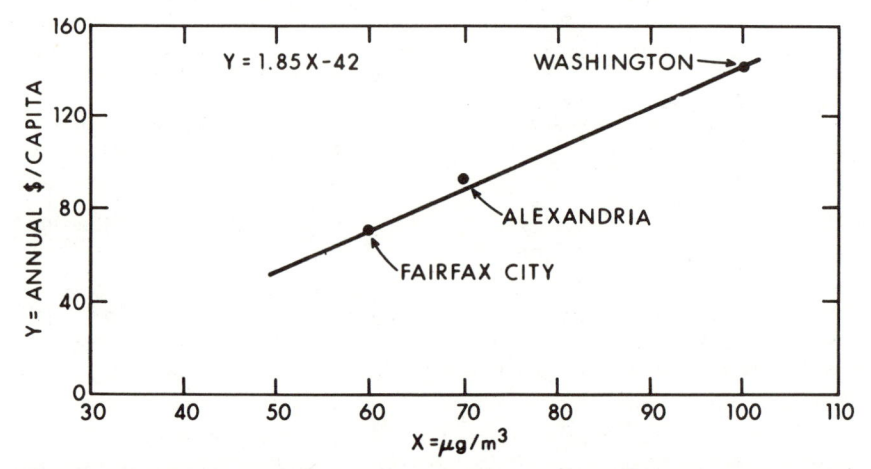

Figure 4.2 Residential soiling costs per capita in the Washington, D.C., area. (Source: R. D. Wilson and D. W. Minnotte, 1969 [22])

levels of particulate pollution concentrations (X) in µg/m³ and the annual cost of damage per capita (Y) in dollars as expressed by

(5) $$Y = 1.85X - 42$$

By simply multiplying the above cost per capita damage function with the total population of an area one can obtain the total damage costs due to residential soiling. Thus the cost of damages caused by pollution for downtown Washington (CP_w) with a population density of 442,000 in 1968 could be calculated for various levels of particulate concentrations (X) according to

(6) $$CP_w = 817{,}700X - 18{,}564{,}000$$

where the slope of the total cost of pollution conveniently gives the costs for each change of 1 µg/m³ in particulate pollution. The value of the slope of the regression line is also called the marginal cost of pollution (MCP). The $MCP_w = 817{,}700$ means that for downtown Washington a reduction of 1 µg/m³ in suspended particulates would save Washingtonians $817,700 per year, whereas an increase in particulate pollution by only 1 µg/m³ would cost them this amount in additional laundry and maintenance costs.

A next step is to calculate the total cost of pollution damages (TCP_{jp}) as a function of an emitter (j) at its present state of control

(p). TCP_{jp} is obtained by multiplying the rate of change of damage per 1 $\mu g/m^3$ for an area (i) given by X_{ijp}. Summation of the multiplications for each of the 32 areas gives the total cost of pollution damages in the Washington area. Thus;

$$(7) \qquad TCP_{jp} = \sum_{i=1}^{32} MCP_i X_{ijp}$$

To obtain finally the net benefits (B_j) from a reduction of particulate emissions as listed in Tables 4.4 and 4.5, one only has to find the difference in damages between the present state of control (p) of an emitter and a higher degree of control (k). Thus,

$$(8) \qquad B_j \text{ state } p \text{ to state } k = TCP_{jp} - \left(\frac{E_{jk}}{E_{jp}}\right) TCP_{jp}$$

In words this equation states that the benefits to be expected from reducing particulate pollution from its present emission level (p) to a lower level (k), the damages attributable to emitter (j) at a state of lesser emissions (k), i.e., (E_{jk}/E_{jp})TCP_{jp}, have to be subtracted from the damages of emitter (j) at the present state of control (p), i.e., TCP_{jp}. The term E stands for particulate emissions from an emitter (j) at its present (p) or some higher state of control (k).

The net benefits shown in Tables 4.4 and 4.5 were calculated from equation (8). In the case of the power plant the maximum net benefit from particulate emission control occurs at the second level of control with a net benefit of $162,000 per year. For the municipal incinerator, however, the maximum net benefits of $1,360,300 per year occur at the maximum control step where only 16 tons of particulates are emitted annually.

For the whole Washington area the total benefits were $22,073,900, and the total air pollution control costs were $2,833,400 resulting in total net benefits from particulate emission control of $19,240,500. So far this whole discussion is an entirely economic one, starting from the premise that an air pollution control system is optimized when the net profits from the control of each pollution emitter are maximized. Health considerations have not yet entered the discussion at all. Also, it remains to be shown what the economic impact would be if not only point but area sources also were controlled.

The example of downtown Washington may clarify these points. At optimum economic control of point sources the annual particulate level could be reduced from 83.3 to 72.9 $\mu g/m^3$. From a public health and welfare point of view, however, it would be desirable to further reduce the particulate pollution level to 60 $\mu g/m^3$ per year. The question now is, Would—even in the absence of a pure monetary profit from control—further control of point rather than that of area sources give the desired reductions? Table 4.6 shows that the point sources prior to control contributed only 11.2 µg/m³ to the total particulate level of 83.3 µg/m³. At optimum particulate pollution control and a reduction to 72,9 µg/m³, the contribution of point sources is only 0.8 µg/m³ to Washington's particulate level. Since natural background pollution (50 µg/m³) is locally not controllable, a further reduction can only be achieved by controlling area sources (22.1 µg/m³) and here in particular the emissions from gasoline and diesel fuels (see Table 4.6).

Table 4.6 Sources of particulate concentrations in downtown Washington, D.C. (µg/m³)

	Present level	Proposed level
Background	50.0	50.0
Area sources	22.1	22.1
Point sources	11.2	0.8
Total	83.3	72.9

Area sources	
anthracite coal	1.1
bituminous coal	0.8
distillate oil	2.6
residual oil	2.0
natural gas	0.8
gasoline	7.9
diesel fuel	4.9
incineration (on site)	1.2
open burning (backyard)	0.0
misc. process sources	0.5
gasoline evaporation	0.0
diesel railroad	0.2
aircraft	0.0
Total	22.1

Source: R. D. Wilson and D. W. Minnotte, 1969 [22].

INCENTIVES FOR AIR POLLUTION CONTROL

Costs of air pollution damages in the United States amount to billions of dollars. Benefits from air pollution control also amount to billions of dollars. It is therefore sound economics to energetically control air pollution. Effective control can be achieved by enforcement of Federal or state regulations or by positive economic incentives. In the case of an active cooperation on the part of the pollutor, the latter procedure is to be preferred. Three incentives for pollution control are usually considered: discharge warrants, effluent fees, and pollution activity taxes.

Discharge warrants would permit the pollutor to emit certain specified amounts of pollution into the atmosphere. Government control agencies would sell these warrants to the highest bidders. Firms would be forced into deciding whether the purchase of pollution control devices with the recovery of important raw materials is not more profitable than buying a license to pollute. The scheme would, however, only work if the warrants would not only cost token amounts but were competitively priced with the control devices.

Under the effluent fees scheme a pollutor would be charged the costs of damages suffered by others as a result of his pollution activity. The idea is to provide to the pollutor an incentive to control his emissions. In reality, however, it would be very costly, and in many cases impossible, to relate a certain pollution damage to a certain pollution source. The additional difficulty of having to assess the real social costs incurred by the damages may render the effluent fee procedure inoperable.

The purpose of the pollution activity tax is to eliminate pollution at its source. The tax has to be high enough so as to make it profitable not to pollute. The incentive in this procedure is the fact that the tax is eliminated if the pollution is. The activity tax can become a sensible pollution control measure if consideration is given not only to economically optimal levels of pollution, but also to social and health costs. The attractiveness of a scheme of emission control at the source lies in the fact that *in situ* cost-damage assessments might become rare, and the enforcement of ambient air quality standards might eventually become unnecessary.

In any case, the determination of an optimum air quality level has to be based primarily on health and social welfare considerations. Cost-benefit analyses are a useful tool to help determine the real costs involved in reaching a healthy air quality.

SUMMARY

It is now generally acknowledged that polluting the nation's atmosphere is a luxury which neither the national economy nor the economy of the individual family can afford any longer. Economists have developed cost-benefit models which show how economic air pollution control is to the pollutor and the nation.

Specifically, this chapter lists individual cost-damage assessments relating to property, traffic, agricultural, and health damages. The real costs of cleaner air (i.e., the costs of the control devices minus the benefits from control) are demonstrated for chemical companies, utility plants, smelters, petroleum refineries, and motor vehicles. The major benefits from air pollution control lie in the sale or reuse of reclaimed effluents, in reduced damage to materials and equipment, and in an improved environment. In reality the general public pays most of the pollution costs: indirectly by cost-sharing the taxes (the public bears 59 percent of the pollutors' costs for control devices), directly by paying higher prices, and above all by suffering from pollution damages. A specific cost-benefit analysis for the Washington, D.C., area demonstrates that for particulate pollution control alone the total net benefits exceed $19 million.

It is finally shown that tax incentives have to be high enough to make it profitable not to pollute the air. Health and social welfare and not mere economic considerations should ultimately determine the optimum air quality.

REFERENCES CITED

[1] H. O. Nourse, "The Effect of Air Pollution on House Values," *Land Economics*, 181-189, May, 1967.
[2] J. D. Williams, et al., *Effects of Air Pollution*, Interstate Air Pollution Study, Phase II Project Dept., USDHEW, PHS, Cincinnati, Ohio, Dec., 1966.
[3] R. G. Ridker, *Economic Costs of Air Pollution—Studies in Measurement*, Frederick A. Praeger, Inc., New York, 1967.
[4] British Transport Commission, London, and BEA, personal communication.
[5] H. Johnson, "The High Cost of Foul Air," reprinted from *The Progressive Farmer*, April 1968.
[6] I. Low, "Smog over the Fields," *New Scientist* 28, 494, November 1968.
[7] D. V. Bates, "Health Costs of Diseases Related to Air Pollution," background paper A4-2-4, Congress on Air Pollution, Montreal, October 31-November 4, 1966.
[8] H. Beaver, "Committee on Air Pollution Report," H.M.S.O. London, 1954, rev. 1958.
[9] *Environmental Pollution*, Rpt. of the Subcommittee on Science, Resource, and Development, U.S. Government Printing Office, Washington, D.C., 1966.
[10] "Air Pollution Versus Materials-Costs," *Materials Protection* 6(5), 47-52, May 1967.
[11] "Solvent Recovery System Saves $39,000 First Year," *Air Engineering* 10(4), 31, July, 1968.
[12] A. Wille and G. H. Weyermuller, "Pollution Control System Saves Plant $60,000," *Chemical Processing* 32(1), 15-20, Jan., 1969.
[13] B. Linsky, "Air Pollution—An Air of Difference," *Transactions 31st North Am. Wildlife. Nat. Res. Conf.*, March 14-16, 1966.
[14] F. A. Ferguson, K. T. Semrau, and D. R. Monti, "SO_2 from Smelters: Byproduct Markets a Powerful Lure," *Env. Science Technology* 4(7), 562-568, July, 1970.
[15] *The Cost of Clean Air*, First Rpt., Secretary HEW to 91st Congress, U.S. Government Printing Office, Washington, D.C., Oct., 1969.
[16] "Gasoline Changes Costly to Refiners," *Oil and Gas J.* 66(52), 34-35, Dec., 1968.
[17] B. Linsky, "Statement before the Subcommittee on Air and Water Pollution of the Senate Committee on Public Works, Washington, D.C., March 18, 1970.
[18] R. D. Wilson and D. W. Minnotte, "Government/Industry Cost Sharing for Air Pollution Control," *JAPCA* 19(10(, 761-766, Oct., 1969 (p. 765 quote).
[19] R. E. Kohn, "Abatement Strategy and Air Quality Standards, in A. Atkisson and R. S. Gaines, eds., *Development of Air Quality Standards*, Merrill Publishing Co., Columbus, Ohio, 1970, pp. 103-122.
[20] W. E. Jackson, et al., "Determining the Costs of Air Pollution Control," *JAPCA* 19(12), 917-923, Dec., 1969.
[21] F. C. Hamburg and F. L. Cross, Jr., "A Training Exercise on Cost-Effectiveness Evaluation of Air Pollution Control Strategies," *JAPCA* 21(2), 66-70, Feb., 1971.
[22] R. D. Wilson and D. W. Minnotte, "A Cost-Benefit Approach to Air Pollution Control," *JAPCA* 19(5), 303-308, May, 1969.

CHAPTER 5

TECHNOLOGY AND AIR POLLUTION

Radioactive Air Pollution

Until the turn of the century atmospheric radioactivity could be attributed to natural sources. Since that time increasing energy demands and the frightening prospect of fossil fuel depletion have forced science and industry to look at another form of energy: nuclear energy. It is ironic that the use of this new "clean" fuel, which produces none of the conventional gases and aerosols, contaminates the environment with an even more insidious poison, namely radioactive pollution. This chapter will not deal with radioactivity per se, which is produced under controlled conditions within reactors or through the use of radioisotopes in scientific research. It will deal with the release of radioactive gases and dusts into the atmosphere, the radioactivity of which is here collectively termed radioactive air pollution.

BASIC TERMINOLOGY

In order to understand the nature of radioactive air pollution a number of basic terms must be explained [1]. The smallest unit of an element, the atom, consists of a nucleus of protons and neutrons which is encircled by electrons. When this nucleus disintegrates or decays, radioactivity is set free. Nuclear emission produces alpha (α) and beta (β) particles and gamma (γ) rays. The slow-moving α-particles can hardly penetrate a thin sheet of paper; the fast-moving β-particles can pass through thin aluminum foil; and the γ-rays, traveling at the speed of light, can penetrate thick metal sheets.

In nature there are elements which are so unstable that they give off radioactivity even without any outside force. Radium (Ra) is such an element, which, by emitting an a-particle along with β- and γ-rays, disintegrates via the unstable radionuclides radon (Rn) and polonium (Po) to the stable atom of lead (Pb). On the other hand, potentially any element can be forcefully converted into another one by rearranging the atomic components. During such a conversion process from unstable to stable elements, enormous amounts of energy are released. On a weight for weight basis, Ra produces 320,000 times more energy than coal [1].

The conversion from one radioactive nucleus (radionuclide) into another one is easiest for atoms that have either only a few or very many protons. Thus, for example, hydrogen (H) with only one proton can be made to combine with other atoms of H or small atoms in a process known as fusion. The energy of the sun, of the stars, and that set free through H-bombs is produced by atomic fusion. If the atom has many protons such as uranium (U) which has 92, then the element is broken into smaller atoms in a process known as fission. The fission process is extensively used in the nuclear energy industry.

Both fission and fusion produce radioactivity which is dangerous because the rays ionize the atoms of any of the substances they penetrate. Ionization causes a chain reaction which damages the penetrated substances, such as the molecules of human cells.

One measure of the amount of radioactive contamination is the Roentgen (r) which is a function of the radioactive energy absorbed in living tissue. A single dental X-ray produces about 1 r and a full mouth X-ray series between 14 and 18 r [2]. Radioactivity can also be expressed in Curie (c) which gives the quantity of radioactive material. For example, the Brookhaven National Center nuclear reactor emits 700 c/hour of radioactive argon-41 (see Table 5.1). Other common measures are millicurie (mc), picocurie (pc), and femtocurie (fc) which are equivalent to 10^{-3}, 10^{-12}, and 10^{-15} c, respectively. Units of measurement related especially to man are rad (radiation absorbed dose), which is used in the same manner as the Roentgen unit, and rem (radiation effective on man), which is a unit of measurement that deals with internal radiation. Rem equals the radiation dose in rad multiplied by the relative biological effectiveness (RBE) of a particular radiation [3] (see Table 5.5). For example, a-particles have a twenty times greater biological effectiveness on the tissue than an equal amount of rads of γ-rays, despite the much greater penetration capability of γ-rays [2].

Radioactive wastes have been dumped into the ocean or buried in thick containers in abandoned mines, but the only way one can ultimately get rid of radioactive contamination is through the disintegration of an isotope until a stable element is produced. The radioactive decay, through which a substance loses half its radioactivity, is a fixed number for each element and is known as the half-life of the substance. As a rule of thumb, the radioactivity of an element has usually disappeared (has reached 0.1 percent of the original value) after an equivalent time period of ten times its half-life [1].

NATURAL RADIOACTIVE AIR POLLUTION

Natural radioactive pollution originates from both radioactive minerals in the earth's crust and from the interaction of cosmic rays with the gases of the atmosphere [4]. Radioactive minerals release, for example, the radioactive noble gases of radon (Rn-222) and thoron (Rn-220), which in turn become absorbed on atmospheric dust charging it with radioactivity.

Radioactivity is also released during the combustion of fossil fuels such as coal and oil. For example, a 1,000 MW (megawatt or 1 million watt) coal-burning power plant with dust collecting equipment would discharge annually about 50 mc of Ra-228 (radium) and 100 mc of Ra-226. An oil-burning power plant of the same capacity with an annual fuel consumption of 460 million gallons would discharge about 0.5 mc of Ra-228 and Ra-226.

Cosmic rays interacting with atmospheric gases produce a number of important radionuclides such as tritium (H-3), carbon (C-14), and beryllium (Be-7). Tritium, the radioactive isotope of hydrogen, exists in the atmosphere in the form of water vapor, and it reaches the earth's surface through rain and snow.

It has become customary to accept natural radioactivity as the norm, and by comparing quantities, to deduce a relative harmlessness of artificially produced radioactivity. Mason [5] brands this as superficial reasoning by pointing out (1) that there is neither a spatial nor temporal uniform distribution of natural radioactivity; (2) that natural radioactivity is neither physiologically nor biologically safe; and (3) that the belief that exposure to "safe" doses above background radiation has no medical foundation, because any amount of increased radiation, no matter how small, will increase the number of persons affected by genetic disease [6].

MAN-MADE RADIOACTIVE AIR POLLUTION

Natural radionuclides from mineral or cosmic sources are basically the same as the artificial or man-made radionuclides produced either by nuclear reactors and plants reprocessing spent reactor fuel or by nuclear and thermonuclear bombs. The flow chart in Figure 5.1 gives the major steps in the processing and use of nuclear energy. Mining, milling, and refining of the uranium and thorium ore may just constitute an occupational hazard to the workers. But the enrichment of uranium in the gaseous diffusion plants for use in nuclear reactors and nuclear weapons may already lead to the release of radioactive gases and dusts. The major threat to the environment from radioactive pollution, however, occurs during the phases of reactor operation, chemical reprocessing of spent fuel, waste disposal, and nuclear weapons testing.

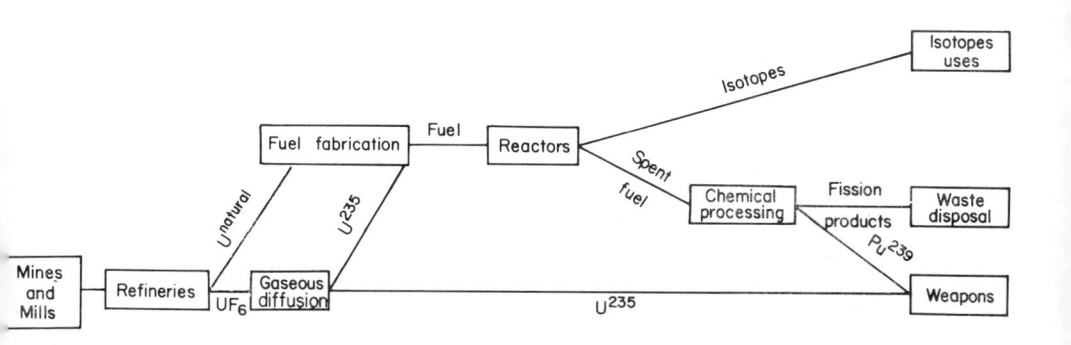

Figure 5.1 Principal steps in the processing and use of nuclear fuel. (Sources: M. Eisenbud, 1963 [3] and 1968 [4])

Nuclear reactors. The three major reactor types in use are the air-cooled, the boiling water, and the pressurized water reactors [8]. Radioactive pollution released from air-cooled reactors consists of argon-41 (Ar-41), discharged fission products produced by defective cladding of fuel rods, and dust in the cooling air activated by neutrons. Boiling water reactors release noble gases such as krypton (Kr) and xenon (Xe) into the steam that is carried to the turbines and discharge fission products into the water through defects in the fuel cladding. The noble gases are released into the atmosphere through the condenser air injector and the stack. Release of radioactive pollution into the atmosphere occurs routinely in 30-minute intervals to allow for the short-lived radionuclides to decay. Pressurized water reactors supply the energy for submarine and surface vessels. The radioactive gases are filtered out from the water coolant, kept in tanks for a few months, and then released to the atmosphere. The lifetime of a reactor is about 30 years.

The amounts of a number of radioactive gases emitted from certain nuclear plants in the United Kingdom and in the United

States are listed in Table 5.1. A typical air-cooled reactor is the one at Brookhaven National Laboratory, which discharges about 750 c/hour of Ar-41 into the atmosphere. Exposure to Ar-41, which has a half-life of 1.8 hours, and which is chemically inert, is controlled by utilizing meteorologically favorable dilution conditions and a 400-foot stack. The 185-MW Dresden nuclear power plant of Morris, Illinois, uses a boiling water reactor. U.S. Atomic Energy Commission regulations permit an annual average discharge of 0.7 c/sec for this reactor [9].

It has been estimated that a 1000-MW nuclear power plant would produce about 13 billion curies of radioactive waste after a year's operation [10]. This is about the same amount of radioactivity that can be released by 14,000 tons of radium. It is more radioactive waste than has been released so far by all atmospheric nuclear weapons tests. The failure in such a system, which is usually referred to as a reactor accident, opens aspects that are indeed frightening.

Reactor accidents. When the heat production is greater than the capacity of the system to cool the reactor, then a serious accident could happen with the inherent release of large amounts of radioactive pollutants. The sudden heat rise could be caused by a malfunction of the moderators which control the rate of fission within the reactor, by chemical reactions among reactor materials, or by a rupture of the piping system with an attendant loss of the coolant.

For the period 1945-1961 the International Atomic Energy Agency enumerates 71 reactor accidents affecting about 123 people [10]. Six people died in four incidents, and in the other incidents a number of people received high dosages of radiation. During the January 1961 accident at the National Reactor Testing Station at Idaho Falls, iodine-131 (I-131) concentrations were measured as far as 100 miles southwest of the station [11].

The most serious reactor accident that has additionally resulted in a considerable contamination of the environment occurred at Windscale in the beautiful Lake District of northwestern England. The accident of October 10, 1957, was caused by a sudden temperature rise and ignition of the uranium fuel cartridges. Of the fission products released to the atmosphere, I-131 was above all the most commonly detected pollutant throughout much of Europe. Along a downwind strip of land about 50 miles long and 10 miles wide in this dairy land milk was destroyed for months. The major radioisotopes and their quantities in curies released during the accident are listed in Table 5.2. The highest I-131 concentration found in a liter of milk was 1.4 mc [12]. The highest I-131 dose in

Table 5.1 Gaseous radioactive wastes from atomic energy plants

Location	Type of plant	Amount of waste	Radioactive content
United Kingdom			
Springfields	feed material production plant	about 1 c/year *	α
Capenhurst	gaseous diffusion plant	about 0.1 c/year	α (uranium)
Calder Hall	nuclear power plant	10 c/hour	argon-41
Chapelcross	nuclear power plant	10 c/hour	argon-41
Dounreay	reactor research center	0.5 mc/hour †	argon-41
Harwell	nuclear research center	30 mc/year	β
		1 mc/year	α
		50 c/hour	argon-41
Amersham	isotope production plant	15 mc/week	iodine-131
Aldermaston	nuclear weapon research center	20 mc/year	β
		3 mc/year	α
United States			
Hanford	plutonium production plant	1 c/day	iodine-131
Idaho	reactor testing station	100,000 c/year	β and noble gases
Oak Ridge National Laboratory	reactor development and chemical processing laboratory	0.25 c/year	α (uranium)
Brookhaven National Laboratory	nuclear research center	700 c/hour	argon-41

*c = curies.   † mc = millicuries.
Source: *Air Conservation*, 1965 [7].

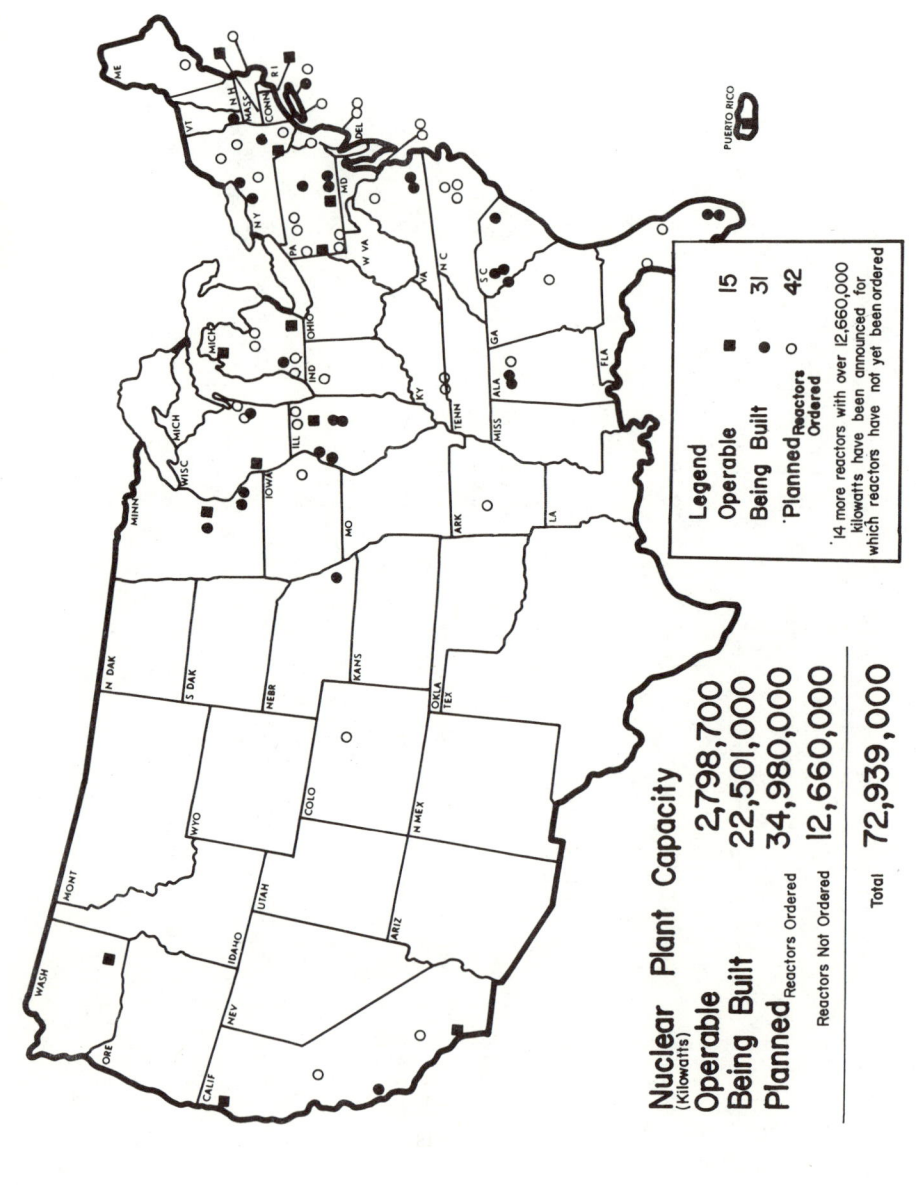

Figure 5.2 Number of nuclear power plants in the United States. (Source: R. Curtis and E. Hogan, 1970 [2])

the thyroid of a child was 19 rad. The Federal Radiation Council suggests protective action at a dose of 30 rad [4].

Table 5.2 Estimated release of radioactivity from the Windscale reactor accident

Radioisotope	Curies
Iodine-131	20,000
Tellurium-132	12,000
Cesium-137	600
Strontium-89	80
Strontium-90	2

Source: United Nations, 1962 [12].

Reactor accidents are extremely serious. With the ever increasing number of nuclear power plants being located in populated areas (Figure 5.2), the probability of reactor accidents increases. Within minutes decisions have to be made as to where the radioactive plume will travel, at what time and at what radioactivity it will arrive over a populated area, and if the population should be evacuated. This is demonstrated by the following hypothetical case which is based on reasonable data and actual calculation procedures.

One second after the accident, a core melt-down of a nuclear reactor releases an estimated 2.5×10^6 c of iodine-131 into the atmosphere of the containment vessel which has an effective cross sectional area of $cA = 200$ m^2. At the recorded wind speed of $u = 6$ m/sec from the southwest the radioactive plume would reach Newtown, the nearest settlement with 5,000 inhabitants, in 16 minutes (the distance $4\sqrt{2} = 5,760$ m divided by 6 m/sec wind speed) (Figure 5.3).

With the help of prepared charts and high-speed computers it can be calculated whether, under the given meteorological conditions, radioactivity at Newtown would exceed the maximum permissible value of 3×10^{-10} μc/ml of I-131 recommended by the Atomic Energy Commission for nonoccupational exposure [3]. At a leak rate of 0.1 percent per day, the leakage of radioactivity from the reactor building per second would be 0.001 day^{-1} /3600 sec hr^{-1} × 24 hr day^{-1} = 1.157×10^{-8}. The source strength Q_I, i.e., the amount of radioactive I-131 that is leaving the building, is equal to the leak rate times the I-131 release. Thus, $Q_I = 1.157 \times 10^{-8}$ sec^{-1} × 2.5×10^6 c = 2.89×10^{-2} c sec^{-1}.

Making use of a modified Gaussian diffusion equation similar to the one presented in Chapter 2, one can calculate the amount of radioactivity (χ) to be expected at Newtown 16 minutes after the accident considering the decay of I-131 (half-life 8.04 days) over this

92 ATMOSPHERIC POLLUTION

short time period to be negligible:

$$\chi(x,y,0,0) = \frac{Q_I}{u(\pi\sigma_y\sigma_z + cA)} \exp\left[-\frac{1}{2}\frac{y^2}{\sigma_y^2}\right]$$

For the distance (reactor to Newtown) of 5,760 m and a partly cloudy windy afternoon in fall with stability class D the diffusion parameters are σ_y = 330 m and σ_z = 96 m [13]. From Figure 5.3, y = 0.5$\sqrt{2}$ = 720 m. Thus, the amount of I-131 to be expected at Newtown would be I-131 = 4.49 x 10^{-9} µc/ml which is fifteen times greater than the maximum permissible value for I-131. Evacuation of Newtown should therefore proceed as fast as possible.

After the population is secured it would be of interest to estimate how long it would take for the calculated concentration of χ_c = 4.49 x 10^{-9} µc/ml to drop below the maximum permissible value for I-131 of χ_{mp} = 3 x 10^{-10} µc/ml. This can be calculated from the decay relationship

$$\chi_{mp} = \chi_c \exp\left(-\frac{0.693t}{L}\right)$$

where L = 8.1 days, the half-life for I-131. Solving for the time t, it can be seen that it takes about 31 days and 9 hours before the I-131 concentration has decayed to the maximum permissible value.

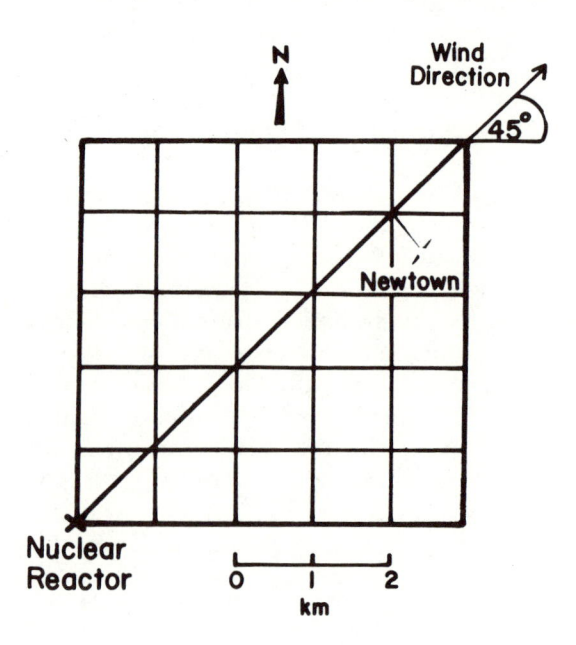

Figure 5.3 Reactor accident.

To demonstrate the principle the simplest possible case with one short-lived radioisotope was shown. It is clear that during a reactor accident a variety of radionuclides with much longer half-lives are released into the air. The evacuation of the 5,000 people of Newtown may already pose a difficult logistic problem. The problems encountered with a city the size of New York would be unimaginable.

Chemical reprocessing of spent fuel. The greatest potential source of radioactive contamination exists when spent reactor fuel is taken to a chemical reprocessing plant. The fuel is dissolved in chemical solutions and the unfissioned uranium and plutonium are separated from the radioactive waste. After 180 days of operation a 500-MW reactor will have produced about 3×10^5 c of both strontium-90 (Sr-90) and cesium-137 (Cs-137) [3]. (Compare this amount of radioactive waste with the values given in Table 5.1.) Since these long-lived radioisotopes hardly decay during the 3 to 12 months they spend at the reprocessing plants, and since they emit gases of Kr-85, Xe-133, and I-131 into the atmosphere they constitute a real threat to the environment [12]. Krypton-85, a noble gas with a half-life of 10.4 years, cannot be removed by chemical or mechanical means and is therefore one of the major nuclear industry radioactive pollutants that is now accumulating in the environment [14]. Eisenbud lists 21 reprocessing centers for the United States alone [3].

Nuclear weapons. Radioactive pollution from nuclear weapons tests is of great concern because of its magnitude and its uncontrolled procedure of release. The number of nuclear explosions in the atmosphere through 1968 together with the detonation power in megatons through 1962 is listed in Table 5.3. (A megaton has the explosive equivalent of a million tons of TNT.)

Table 5.3 Number of nuclear explosions in the atmosphere, 1945-1968, and approximate total yield in megatons of all nuclear weapons tests through 1962

Country	Testing period	Number of tests	Testing period	Megatons
USA	1945-1958	174	1945-1951	1
USSR	1949-1961	106	1952-1954	60
UK	1952-1958	21	1955-1956	28
France	1960-1968	15	1957-1958	85
China	1964-1968	8	1961	120
			1962	217
Total		324	Total	511

Sources: M. Eisenbud, 1963 [3]; and 1968 [4]; C. E. Junge, 1963 [15]; B. Shleien, et al., 1970 [16].

A large number of articles has been published on the various aspects of nuclear weapons tests. Some of these deal with the dispersion and transport of radioactive material in the atmosphere [17]; radioactive fallout from underground tests [18]; the Radiological Health Data Reports with data of β-radioactivity for the United States, Canada, Mexico, and South America [19]; or more locally for AEC nuclear plants at Paducah, Kentucky, or Piketon, Ohio [20]. The nuclear weapons tests in the South Pacific have provoked the special vigilance of Australian scientists [21]. Since the adoption of the atmospheric test ban treaty by the United States, the United Kingdom, and the Soviet Union in 1963, vigilance has concentrated on French and Chinese weapons tests. France and China have not yet signed the treaty [22]. Dyer and Hicks [21] have shown that each period of nuclear tests is followed by a sharp increase in β-radioactivity (Figure 5.4).

Of all sources, nuclear weapons tests release by far the greatest amounts of radioactive pollution. For example, a 1-megaton bomb produces within one minute after detonation about 4×10^{12} c of γ-activity. (Compare with figures in Tables 5.1.) There are in general three types of fallout from a nuclear explosion: (1) large particles (> 50 μm) which are deposited after several days within a few hundred miles of the test area; (2) smaller particles that are carried around the globe within the troposphere for several weeks until they are eventually washed out by rainfall; and (3) small particles that have been injected into the stratosphere where they can remain for years before they finally reach the ground.

It has been estimated that about one-third of all the fission material that has been injected into the atmosphere has fallen out in the immediate vicinity of the test area [4]. However, of the remaining two-thirds, 90 percent were injected into the stratosphere and only 10 percent into the troposphere. It is perhaps blasphemous to offer in this deadly context some consolation; but the higher these substances are emitted into the atmosphere, the longer they remain there, and the weaker will be their radioactivity when they return to the earth's surface.

The amount of β-radioactivity for the United States in 1966 is shown in Table 5.4. States grouped around and downwind of the Nevada test sites have the highest concentrations. This might perhaps be surprising so many years after the 1963 atmospheric test ban. Underground tests have not been banned, however, and subsurface tests still vent a good deal of radioactivity into the atmosphere through cracks in the ground, etc.

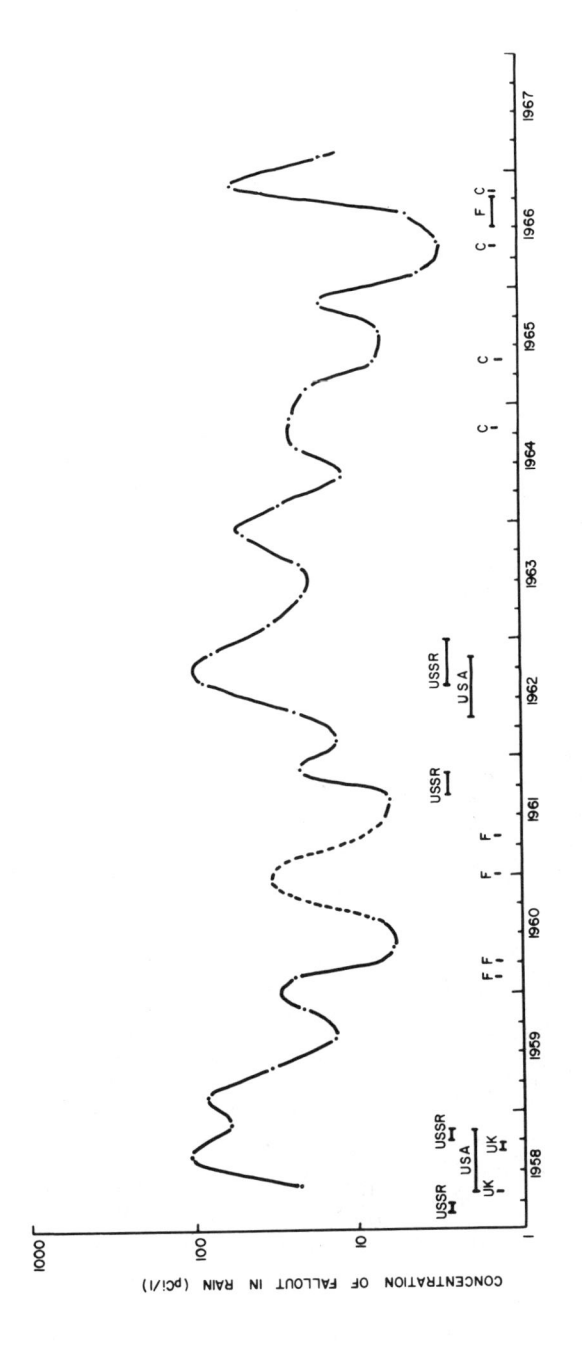

Figure 5.4 Nuclear weapons tests and fallout of long-lived β-emitters at Aspendale, Australia, since 1958. (Source: A. J. Dyer and B. S. Hicks, 1967 [21])

Table 5.4 Beta radioactivity in the United States

Rank	State	Total (pCi/m^3)	Rank	State	Total (pCi/m^3)
1	Nevada	0.55	24	Utah	0.23
2	Arizona	0.53	25	Kansas	0.22
3	Texas	0.42	26	Washington, D.C.	0.21
4	Colorado	0.40	27	South Dakota	0.21
5	Wyoming	0.38	28	West Virginia	0.21
6	Michigan	0.35	29	California	0.20
7	Missouri	0.35	30	Nebraska	0.20
8	New Mexico	0.33	31	New Hampshire	0.20
9	Wisconsin	0.33	32	North Carolina	0.20
10	Illinois	0.30	33	Rhode Island	0.20
11	Ohio	0.30	34	South Carolina	0.20
12	Oregon	0.30	35	Virginia	0.20
13	Kentucky	0.29	36	Iowa	0.19
14	Oklahoma	0.29	37	Pennsylvania	0.19
15	Arkansas	0.28	38	Maryland	0.17
16	Louisiana	0.28	39	Montana	0.17
17	Idaho	0.26	40	New Jersey	0.17
18	Tennessee	0.26	41	Vermont	0.17
19	Delaware	0.25	42	Connecticut	0.16
20	Mississippi	0.25	43	Maine	0.16
21	New York	0.25	44	Minnesota	0.15
22	Alabama	0.23	45	Georgia	0.13
23	Indiana	0.23	46	Washington	0.13

Source: DHEW, PHS, 1966 [23].

Commercial use of nuclear explosives. For quite some time nuclear explosives in mining and engineering projects have gone largely unnoticed. In December 1967 Project Gasbuggy was started in New Mexico. The idea was to fracture rocks with nuclear explosives and hopefully to free trapped oil or gas. The explosion set free Kr-85 and H-3, the latter becoming incorporated into the molecules of the methane and thus making the gas radioactive. In order to get rid of the radioactivity it has been necessary to burn or flare the gas-well for a year and a half. Already about 200 million cubic feet of this nonrenewable resource have been needlessly wasted, and the radioactivity is still too large for commercial use. Project Bronco has been devised to free oil trapped in oil-shale formations, and Project Sloop, to extract copper from low-grade copper ores.

Nuclear explosives have also been suggested for highway and railroad construction, right-of-way through mountainous terrain, construction of dams and rockfill structures. Project Plowshare, the construction of a sea-level canal in Panama, has been discussed since 1962 [24]. Apart from radioactive hazards to the population and the radioactive contamination of the debris within the excavation zone, a violent disruption of local ecosystems is also expected when a direct contact of the different Pacific and Atlantic biological species occurs. In any case, all of these subsurface explosions contribute to radioactive contamination of the atmosphere; the amount depends on the type, magnitude, and depth of the placed explosive.

EFFECT OF RADIOACTIVE AIR POLLUTION ON MAN

Exposure to radioactive pollution produces both somatic effects which can be seen in people living now, and genetic effects which are passed on to their descendants. Whole-body exposure to radiation is due to natural radiation and would be experienced in nuclear weapons tests and reactor accidents. Therapeutical radiation is usually only administered to parts of the body.

Acute or short-term effects due to radioactive pollution consist of radiation burns, nausea, changes in the blood, and disorders of the intestines and the central nervous system [1]. Latent and chronic or long-term effects include a variety of cancers, impairment of growth and development, shortening of life-span, and genetic disorders [7].

Radionuclides affect man mainly through background radiation, through inhalation, and through intake via the food chain. In terms of their effects on man the following four radionuclides are of greatest concern.

Iodine-131 (I-131). A β-γ-emitter with a half-life of 8.1 days, this short-lived nuclide affects the body internally and externally. It is abundantly produced in fission processes, and via the cattle-milk food chain is deposited in the thyroid glands. Cancer of the thyroid has been reported to occur in children after exposures as low as 150 rem.

Strontium-90 (Sr-90). A β-emitter with a half-life of about 28 years, Sr-90 is one of the most dangerous radioactive pollutants. Because of its chemical similarity to calcium (Ca) it is deposited directly in the skeleton. Again via the food chain vegetation, milk, and meat, Sr-90 reaches man. Apart from the somatic effects such as leukemia and bone cancer, Sr-90 has been reported to have significant genetic effects on mice [7].

Cesium-137 (Cs-137). A β-γ-emitter with a half-life of about 30 years, Cs-137 is deposited mainly in the soft tissue and irradiates the whole body. Fortunately, it is eliminated from the body in a biological half-life of about 100 days [1]. Its γ-rays are a genetic hazard because they irradiate the gonads.

Carbon-14 (C-14). A β-γ-emitter with a half-life of about 5,730 years, C-14 is readily absorbed by the body leading to whole-body radiation and to both somatic and genetic adverse effects. Nuclear weapons tests have produced large amounts of C-14 resulting in an 80 percent increase over the natural C-14 level by summer 1963 [7].

The effects of delayed radiation on atomic bomb survivors have been studied intensively [25]. Over the period 1948-1953, 71, 280 pregnancies were examined in Hiroshima and Nagasaki. It was found that certain radiation doses calculated for certain distances away from the hypocenter correlate significantly with radiation-induced disorders and diseases such as abnormalities in cell development, chromosomal damage, leukemia, thyroid, lung and breast cancer, impairment of growth, and increase in fetal, infant, and general mortality.

The Federal Radiation Council and the United Nations have issued radiation protection guides which are, however, not binding on federal agencies. The threshold values given in Table 5.5 for whole

Table 5.5 Radiation protection guides (RPG) of the Federal Radiation Council and natural background radiation given by the United Nations Scientific Committee on the Effects of Atomic Radiation

Tissue or organ	RPG for individuals	RPG for a suitable sample of the exposed population	Natural background radiation
Whole body	0.5 rem/year	0.17 rem/year	
Gonads		5 rem/year	0.125 rem/year
Thyroid	1.5 rem/year	0.5 rem/year	
Bone Marrow	0.5 rem/year	0.17 rem/year	0.122 rem/year
Bone	1.5 rem/year	0.5 rem/year	0.137 rem/year

Source: *Air Conservation*, 1965 [7].

body, gonads, and bone marrow are $1^{1/3}$, for bone $2^{2/3}$, and for thyroid exposure 4 times greater than the natural background radiation. Obviously one allows here for additional exposure to man-made radioactive pollution, although any amount, no matter how small, will affect people through genetic diseases [6].

SUMMARY

The release of radioactive gases and dusts into the atmosphere produces alpha and beta particles and gamma rays, here collectively termed radioactive air pollution. Ionization from fission and fusion processes destroys the molecules of human cells. Radioactive contamination can be measured in Roentgen (r), Curie (c), radiation absorbed dose (rad), and radiation effective on man(rem). As a rule of thumb, the radioactivity of an element has usually disappeared after an equivalent time period of 10 times its half-life.

Natural radioactive air pollution originates both from radioactive minerals in the earth's crust and from the interaction of cosmic rays with the gases of the atmosphere. Man-made radioactive air pollution is produced either by nuclear reactors and plants reprocessing spent reactor fuel, or by nuclear weapons tests and the commercial use of nuclear explosives. With the ever increasing number of nuclear reactors the prospect of reactor accidents increases. An example demonstrates the calculation of radioactivity and evacuation procedures after a hypothetical reactor accident.

Exposure to radioactive pollution produces both somatic effects which can be studied in people living now, and genetic effects which are passed on to future generations. The four radionuclides, iodine-131, strontium-90, cesium-137, and carbon-14, which have the greatest adverse effects on man, are discussed. The belief that "safe" doses exist above background radiation has no medical foundation, because any amount of increased radiation, no matter how small, will increase the number of persons affected by genetic disease.

REFERENCES CITED

[1] *Air Pollution Primer*, National Tuberculosis and Respiratory Disease Association, New York., 1969, pp. 46-54.
[2] R. Curtis and E. Hogan, *Perils of the Peaceful Atom*, Ballantine Books, New York, 1970, pp. 168 ff.
[3] M. Eisenbud, *Environmental Radioactivity*, McGraw-Hill Book Company, New York, 1963, pp. 270 ff.
[4] _____, "Sources of Radioactive Pollution," in A. C. Stern, *Air Pollution*, 2nd ed., vol. 1, Academic Press, Inc., New York, 1968, pp. 121-147.
[5] P. F. Mason, "Spacial Variability of Atmospheric Radioactivity in the U.S.," *Proceedings of the Association of American Geographers*, vol. 2, 1970, pp. 92-97.
[6] A. M. O. Veale, "Biological Effects of Fallout," *New Zealand Science Review*, 24(4), 49-50, 1966.
[7] *Air Conservation*, Report Air Conservation Commission of the AAAS, Publ. No. 80, AAAS, Washington, D.C., 1965, pp. 158-194.
[8] J. G. Terrill, Jr., et al., "Environmental Aspects of Nuclear and Conventional Power Plants," *Ind. Med. and Surgery* 36, 412-419, June, 1967.
[9] J. C. Carroll and J. O. Schuyler, "The Humboldt Bay Reactor Operating Experience," *Nuclear Safety* 6, 441-451, 467, 1965.
[10] H. B. Smets, "A Review of Nuclear Reactor Incidents," in *Reactor Safety and Hazards Evaluation Techniques*, vol. 1, *Proceedings of a Symposium*, Vienna, 1962, pp. 89-110.
[11] J. R. Horon and W. P. Gammill, "The Health Physics Aspects of the SL-1 Accident," *Health Physics* 9, 117-186, 1963.
[12] United Nations, *General Assembly Report of the UN Scientific Committee on the Effects of Atomic Radiation*, 17th Session, Suppl. No. 16 (A/5216), United Nations, New York, 1962.
[13] D. B. Turner, *Workbook of Atmospheric Dispersion Estimates*, USDHEW, PHS, Cincinnati, Ohio, 1967, revised 1969.
[14] R. Liberace and J. R. Coleman, "Nuclear Power Production and Estimated Kr-85 Levels," *Radiol. Health Data Report* 7, 615-621, Nov., 1966.
[15] C. E. Junge, *Air Chemistry and Radioactivity*, Academic Press, Inc., New York, 1963, pp. 239, 242.
[16] B. Shleien, et al., "Strontium-90, Strontium-89, Plutonium-239, and Plutonium-238 Concentrations in Ground-Level Air, 1964-1969," *Env. Science Technology* 4(7), 598-602, July, 1970.
[17] E. F. Danielsen, "Radioactivity Transport from Stratosphere to Troposphere," *Mineral Industries* 33(6), 1-8, Mar., 1964.
[18] E. A. Martell, "Iodine-131 Fallout from Underground Tests II," *Science* 148(3678), 1756-1757, June, 1965.
[19] Anon., "Radioactivity in Airborne Particulates and Precipitation," *Radiol. Health Data Report* 11, 85-93, Feb., 1970.
[20] Anon., "Environmental Levels of Radioactivity at Atomic Energy Commission Installations," *Radiol. Health Data Report* 9, 56-61, Jan., 1968.

[21] A. J. Dyer and B. B. Hicks, "Radioactive Fallout from the French 1966 Pacific Tests," *Australian J. Science* 30(5), 168-170, Nov., 1967.
[22] D. L. Swindle and P. K. Kuroda, "Variation of the Sr-89/Sr-90 Ratio in Rain Caused by the Chinese Nuclear Explosions of December 28, 1966, and June 17, 1967," *J. Geophys. Res.* 74(8), 2136-2140, Apr., 1969.
[23] USDHEW, PHS, *Air Quality Data*, NAPCA, Publ. No. APTD 68-9, 1966.
[24] E. A. Martell, "Plowing a Nuclear Farrow," *Environment* 11(3), 2-10, 12-13, 26-28, Apr., 1969.
[25] T. Hashizume, et al., "Estimation of the Air Dose from the Atomic Bombs in Hiroshima and Nagasaki," *Health Physics* 13, 149-161, Feb., 1967.

CHAPTER 6

MEASURES OF AIR POLLUTION CONTROL

Air Pollution Legislation

GUIDING PRINCIPLES FOR AIR POLLUTION LEGISLATION

Basically the objective of any air pollution legislation should be to maintain air as clean as possible, but not dirtier than a specified quality [1]. More specifically, the goals should be (1) to preserve the health and welfare of man; (2) to protect plant and animal life; (3) to prevent any damage to physical property; (4) to ensure visibility for safe air and ground transportation; and (5) to guarantee an aesthetically pleasing environment [2].

Legal authority to control local air pollution problems is usually vested in municipal, county, and state agencies. The Federal government, however, has often found itself compelled to take action when either a local agency failed to act, or when local authority was nonexistent. Legal action to control air pollution should be guided by the following axioms [3] : (1) Control actions should not be postponed until an air pollution disaster has occurred. (2) Since it is more economical to control air pollutants at their sources, prevention of pollution emission rather than establishment of costly monitoring and warning systems should be the prime control criterion. (3) Effective air pollution control can only be accomplished if local control agencies and the public closely cooperate in implementing and enforcing stringent air quality standards which insure protection of the public from adverse effects.

It is thus clear that legislation controlling air pollution is necessary. Such legislation, however, will be ineffective unless it is enforced. Enforcement requires constant public awareness, public supervision, and public pressure. The public has the right and the duty to enforce legislation that has been designed not only to maintain but also to improve the air quality of an urban area in spite of continued urbanization and economic growth. Public vigilance is of utmost importance in enforcing legislation that specifies that there shall be no deterioration of air quality in relatively clean areas.

Legislative texts often tend to be vague and thus not readily enforcible. The following four examples, the first two from the Air Quality Act of November 4, 1967 [4], and the last two from the Clean Air Amendments of December 1970 [5] illustrate the use of "buzz words" [author's italics]:

... protection of health should be considered a *minimum requirement* ...

or ... *wherever possible* ... standards should be established which enhance the quality of the environment. And

National primary ambient air quality standards ... [must allow] an *adequate margin* of safety, ...

or ... national secondary ambient air quality standards [must be] attained within a *reasonable period of time* ...

H. C. Wohlers, et al [1], point out that the buzz words may sound nice but that they are unsuitable for enforcement, and they ask,

where would the USA be today if our forefathers prepared the Preamble of the Constitution in the following buzz word manner:

"We, the people of the United States, in order to form a *slightly* more perfect union, establish *reasonable* justice, insure *limited* domestic tranquility, provide for a *level of* common defense, promote general welfare *under specified conditions* and *attempt* to secure the *possible* blessings of liberties to ourselves and to a *portion of our* posterity, do ordain and establish this Constitution for the United States of America."

HISTORICAL REVIEW OF AIR POLLUTION LEGISLATION

In 1306, during the reign of Edward I of England, the first smoke abatement law prohibited the use of "sea coal" because of its deleterious effects to health [6]. In 1307 a violator of this law was executed [2]. Apparently in later years prosecutors became more humane, but the inhumane use of the atmosphere as a giant sewer continued unabated. Throughout the following centuries foreign policy distracted from the intolerable sanitation conditions at home. It was not until 1863 when the world's first comprehensive clean air act, the Alkali Act, was passed in Britain. This act was designed to control emissions of offensive gases, smoke, grit, and dust from specified industries [2].

In the United States the first Federal air pollution control program was introduced in 1955. The major components of the act stipulated that the Federal government would provide research and technical assistance, but that states and local governments would be responsible for pollution control at the source [3]. Since 1960 Federal technical assistance and training has been provided under the auspices of the Department of Health, Education, and Welfare (HEW) through the Division of Air Pollution, later renamed the Center for Air Pollution Control and National Air Pollution Control Administration. In 1970 the agency was renamed Air Pollution Control Office (APCO), but operating now within the newly established Environmental Protection Agency (EPA).

In part echoing the 1955 Clean Air Act, the Air Quality Act of November 1967 promulgated for the first time the acceptance of responsibility by the Federal government if local control agencies failed. The major objectives of the Air Quality Act of 1967 were "to protect and enhance the quality of the Nation's air resources so as to promote the public health and welfare and the productive capacity of its population," and "to insure that air pollution problems will in the future, be controlled in a systematic way" [7]. The act also promulgated that HEW should designate air quality control regions (Table 6.1), develop air quality criteria and control techniques, supervise the establishment of local air quality standards, and approve local implementation plans.

THE CLEAN AIR AMENDMENTS OF 1970

Major changes. On December 30, 1970, the president of the United States signed into law the Clean Air Amendments of 1970 "to amend the Clean Air Act [of 1967] and to provide for a more effective program to improve the quality of the Nation's air" [5]. Because of the enormous impact the amendments will have on the nation's air quality throughout the next decade, they are translated into flow charts for control action with corresponding time tables for stationary and mobile sources (Figures 6.1 and 6.2).

The following are comments on the major deviations from the 1967 Air Quality Act: (1) The new amendments prescribe that any new air quality control regions will have to be designated by April 1, 1971. This imposes a considerable time pressure on EPA so that they will presumably accept boundaries recommended by state officials and not by a "Consultation Report." (2) The definition of adverse effects is much broader under the 1970 act, encompassing effects on visibility and climate, on economic values, and on personal comfort and well-being. (3) One of the major changes is the setting of national primary and secondary air quality standards which must protect the public health and welfare from known or anticipated adverse effects and must provide an adequate margin of safety. The national standards, however, are set at an intermediate level without consideration for the propagated margin of safety. It rather appears that the standards are set with the hope of achieving a certain amount of air pollution control in heavily polluted areas without preventing air quality deterioration in relatively clean areas. States, it is true, have the option to set their own more stringent standards. Past experience teaches, however, that states, local governments, and corporations, due to the nature of their interests, have only reluctantly, and then only when forced to do so, adopted standards that the public wanted. (4) With the Federal government setting upper limits, logically public hearings on local air quality standards are no longer mandatory. Under the 1970 act public control and public hearings are shifted to the later implementation and enforcement stages. What air quality does the public enforce if the state can adopt standards far above presently existing levels? What air quality can the public expect if, with all the legally granted extension, air quality standards do not have to be enforced before June 1, 1977 (see Figure 6.1)? It has been suggested that the Clean Air Amendments of 1970 have the potential to make the United States a nation of equal pollutees. Only vigilance and deep involvement of the public can prevent this.

Critique regarding stationary sources. For each pollutant EPA may establish a consulting committee of technically qualified representatives of state and local government, industry, and the academic fields. Here no provision is made to consult the citizenry. Local agencies in areas in which the present air quality levels are below the published levels of adverse effects may, but are not required to, adopt more stringent standards. Secondary air quality standards have to be attained within a "reasonable" period of time. Such a rubber time clause is not enforceable.

It is not clear what happens when EPA disapproves of portions of an implementation plan. Does the portion approved go into effect? Further, it is unclear whether EPA or a local agency has the authority to enforce performance standards in Federal facilities. EPA is required neither to publish guidelines for minimum monitoring systems and personnel requirements, nor to consider the synergistic effects of hazardous pollution emissions.

When court action is taken by EPA there is no provision for a public hearing or a public inspection of records of any such sessions. Civil action should be brought before courts wihin whose jurisdiction the offense occurs. The statement, "any person knowlingly violating

ATMOSPHERIC POLLUTION

Table 6.1 Air quality control regions in the United States

Region / State(s)	Region / State(s)	Region / State(s)

1. Washington, D.C.
   District of Columbia
   Maryland
   Virginia
	18. Indianapolis
	    Indiana
		36. Salt Lake City
		    Utah

2. New York City
   Connecticut
   New Jersey
   New York
	19. Minneapolis - St. Paul
	    Minnesota
		37. New Orleans
		    Louisiana

	20. Milwaukee
	    Wisconsin
		38. Miami
		    Florida

3. Chicago
   Illinois
   Indiana
	21. Providence
	    Massachusetts
	    Rhode Island
		39. Oklahoma City
		    Oklahoma

	22. Seattle - Tacoma
	    Washington
		40. Omaha
		    Iowa
		    Nebraska

4. Philadelphia
   Delaware
   New Jersey
   Pennsylvania
	23. Louisville
	    Indiana
	    Kentucky
		41. Honolulu
		    Hawaii

5. Denver
   Colorado
	24. Dayton
	    Ohio
		42. Beaumont - Port Arthur
		    Texas

6. Los Angeles
   California
	25. Phoenix
	    Arizona
		43. Charlotte
		    North Carolina
		    South Carolina

7. St. Louis
   Illinois
   Missouri
	26. Houston
	    Texas
		44. Portland
		    Maine

	27. Dallas - Ft. Worth
	    Texas
		45. Albuquerque
		    New Mexico

8. Boston
   Massachusetts
	28. San Antonio
	    Texas
		46. Lawrence-Lowell-Manchester
		    Massachusetts
		    New Hampshire

9. Cincinnati
   Indiana
   Kentucky
   Ohio
	29. Birmingham
	    Alabama	

	30. Toledo
	    Michigan
	    Ohio
		47. El Paso
		    Texas
		    New Mexico

10. San Francisco
    California
	31. Steubenville
	    Ohio
	    West Virginia
		48. Las Vegas
		    Nevada
		    New Mexico

11. Cleveland Ohio		

12. Pittsburgh
    Pennsylvania
	32. Chattanooga
	    Georgia
	    Tennessee
		49. Fargo-Moorhead
		    Minnesota
		    North Dakota

13. Buffalo New York		

14. Kansas City
    Kansas
    Missouri
	33. Atlanta
	    Georgia
		50. Boise
		    Idaho

	34. Memphis
	    Arkansas
	    Mississippi
	    Tennessee
		51. Billings
		    Montana

15. Detroit
    Michigan
		52. Sioux Falls
		    South Dakota

16. Baltimore Maryland		

17. Hartford - Springfield
    Connecticut
    Massachusetts
	35. Portland
	    Oregon
	    Washington
		53. Cheyenne
		    Wyoming

		54. Anchorage Alaska

(continued)

Table 6.1 (continued)

Region State(s)	Region State(s)	Region State(s)
55. Burlington 　　Vermont 　　New York	69. Joplin-Fayetteville 　　Missouri 　　Oklahoma 　　Kansas 　　Arkansas	81. Douglas-Lordsburg 　　Arizona 　　New Mexico
56. San Juan 　　Puerto Rico		82. Dubuque 　　Iowa 　　Illinois 　　Wisconsin
57. Virgin Islands	70. La Crosse-Winona 　　Wisconsin 　　Minnesota	
58. Allentown-Bethlehem- 　　Easton-Phillipsburgh 　　Pennsylvania 　　New Jersey	71. Menominee-Escanaba- 　　Marinette 　　Michigan 　　Wisconsin	83. Keokuk 　　Iowa 　　Missouri 　　Illinois
59. Binghamton 　　New York 　　Pennsylvania	72. Mobile-Pensacola- 　　Gulfport 　　Alabama 　　Florida 　　Mississippi	84. Lewiston-Moscow- 　　Clarkston-Pullman 　　Idaho 　　Washington
60. Bristol-Johnson- 　　Kingsport 　　Virginia 　　Tennessee		85. Norfolk-Elizabeth City 　　Virginia 　　North Carolina
61. Columbus-Phenix City 　　Georgia 　　Alabama	73. Paducah-Cairo 　　Kentucky 　　Illinois	86. Savannah-Beaufort 　　Georgia 　　South Carolina
62. Cumberland-Keyser 　　Maryland 　　West Virginia	74. Parkersburg-Marietta 　　West Virginia 　　Ohio	87. Shreveport-Texarkana 　　Louisiana 　　Texas 　　Arkansas
63. Duluth-Superior 　　Minnesota 　　Wisconsin	75. Rockford-Janesville- 　　Beloit 　　Illinois 　　Wisconsin	88. Sioux City 　　Iowa 　　Nebraska
64. Youngstown-Erie 　　Ohio 　　Pennsylvania	76. Scottsboro-Jasper 　　Alabama 　　Tennessee	89. Spokane-Coeur d'Alene 　　Washington 　　Idaho
65. Evansville-Owensboro- 　　Henderson 　　Indiana 　　Kentucky	77. South Bend-Elkhart- 　　Benton Harbor 　　Indiana 　　Michigan	90. Vicksburg-Tallulah 　　Mississippi 　　Louisiana
66. Florence-Corinth 　　Alabama 　　Mississippi 　　Tennessee	78. Augusta-Aiken 　　Georgia 　　South Carolina	91. Tulsa 　　Oklahoma
67. Fort Smith-Muskogee 　　Arkansas 　　Oklahoma	79. Berlin-Rumford 　　New Hampshire 　　Maine	
68. Huntington-Ashland- 　　Portsmouth-Ironton 　　West Virginia 　　Kentucky 　　Ohio	80. Davenport-Rock Island- 　　Moline 　　Iowa 　　Illinois	

Source: The Conservation Foundation, Aug., 1970 [8].

108 ATMOSPHERIC POLLUTION

Figure 6.1 Control of air pollution and time table for stationary sources under the Clean Air Amendments of December, 1970. (Source: Public Law 91-1783 [5])

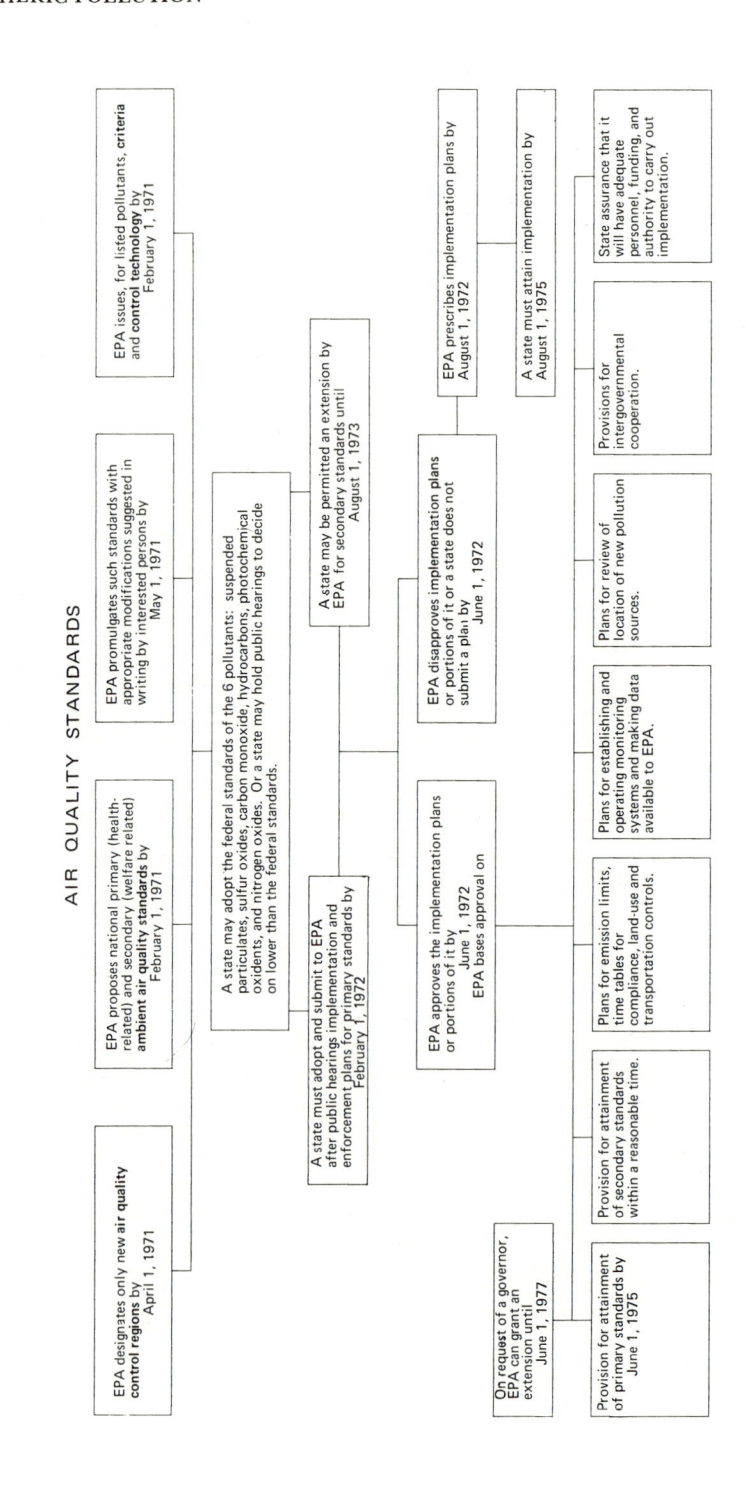

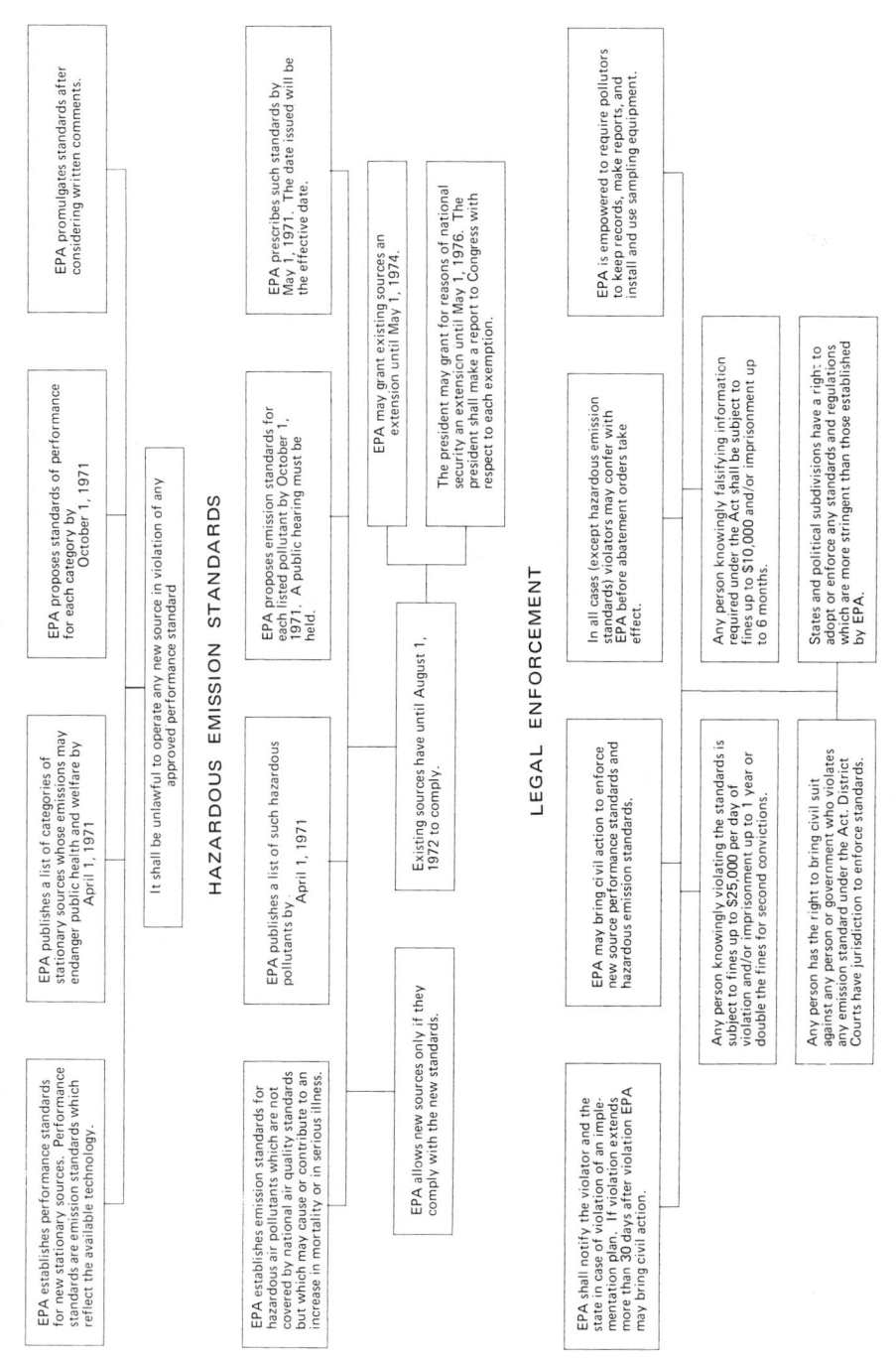

MEASURES OF AIR POLLUTION CONTROL 109

PERFORMANCE STANDARDS

- EPA establishes performance standards for new stationary sources. Performance standards are emission standards which reflect the available technology.
- EPA publishes a list of categories of stationary sources whose emissions may endanger public health and welfare by April 1, 1971.
- EPA proposes standards of performance for each category by October 1, 1971.
- EPA promulgates standards after considering written comments.
- It shall be unlawful to operate any new source in violation of any approved performance standard.

HAZARDOUS EMISSION STANDARDS

- EPA establishes emission standards for hazardous air pollutants which are not covered by national air quality standards but which may cause or contribute to an increase in mortality or in serious illness.
- EPA publishes a list of such hazardous pollutants by April 1, 1971.
- EPA proposes emission standards for each listed pollutant by October 1, 1971. A public hearing must be held.
- EPA prescribes such standards by May 1, 1971. The date issued will be the effective date.
- EPA allows new sources only if they comply with the new standards.
- Existing sources have until August 1, 1972 to comply.
- EPA may grant existing sources an extension until May 1, 1974.
- The president may grant for reasons of national security an extension until May 1, 1976. The president shall make a report to Congress with respect to each exemption.

LEGAL ENFORCEMENT

- EPA may bring civil action to enforce new source performance standards and hazardous emission standards.
- In all cases (except hazardous emission standards) violators may confer with EPA before abatement orders take effect.
- Any person knowingly falsifying information required under the Act shall be subject to fines up to $10,000 and/or imprisonment up to 6 months.
- EPA is empowered to require pollutors to keep records, make reports, and install and use sampling equipment.
- EPA shall notify the violator and the state in case of violation of an implementation plan. If violation extends more than 30 days after violation EPA may bring civil action.
- Any person knowingly violating the standards is subject to fines up to $25,000 per day of violation and/or imprisonment up to 1 year or double the fines for second convictions.
- Any person has the right to bring civil suit against any person or government who violates any emission standard under the Act. District Courts have jurisdiction to enforce standards.
- States and political subdivisions have a right to adopt or enforce any standards and regulations which are more stringent than those established by EPA.

110 ATMOSPHERIC POLLUTION

Figure 6.2 Control of air pollution and time table for mobile sources under the Clean Air Amendments of December, 1970. (Source: Public Law 91-1783 [5])

MOTOR VEHICLE EMISSION STANDARDS

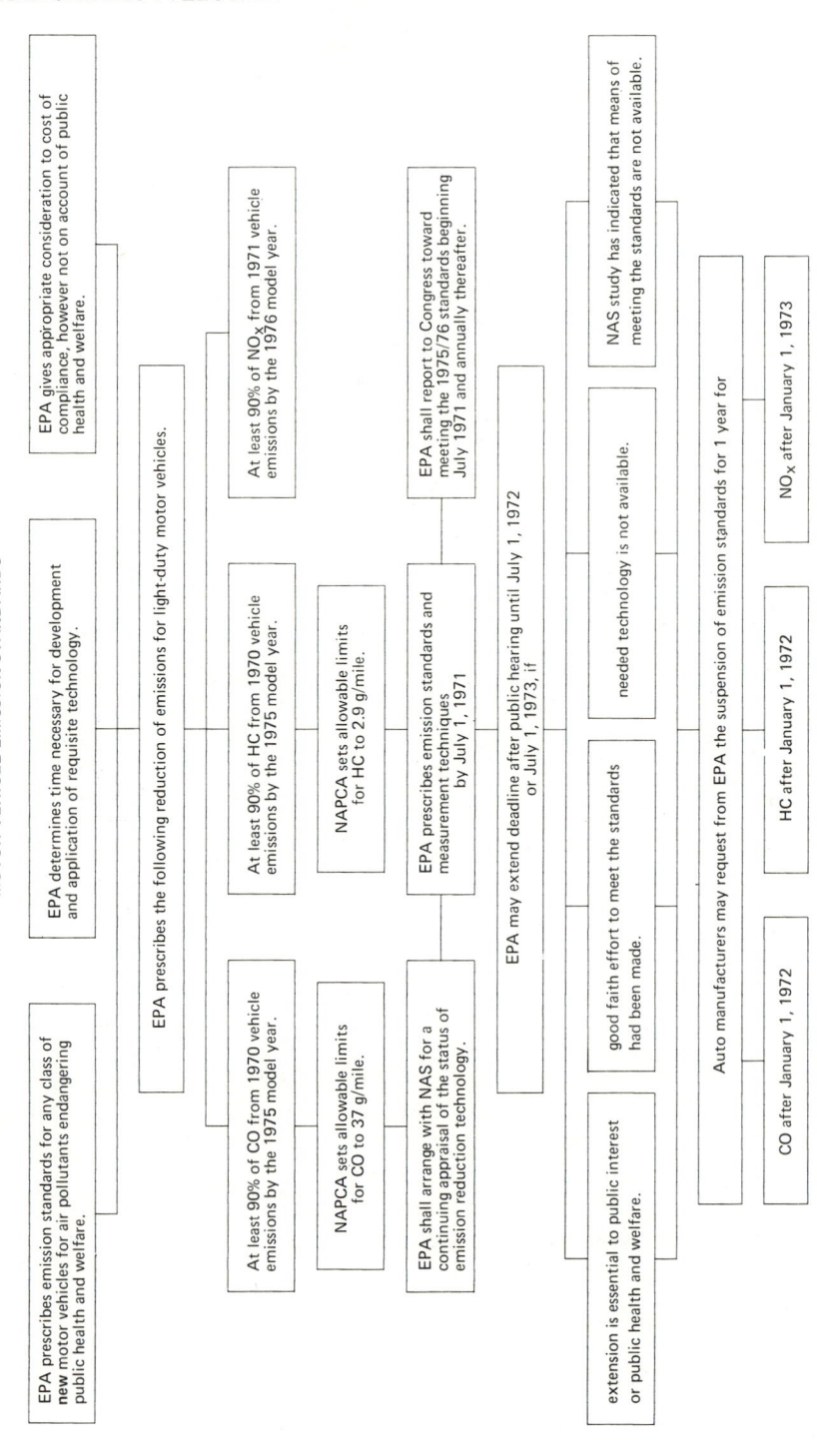

COMPLIANCE TESTING AND CERTIFICATION, REGULATION OF FUELS AND ADDITIVES ENFORCEMENT

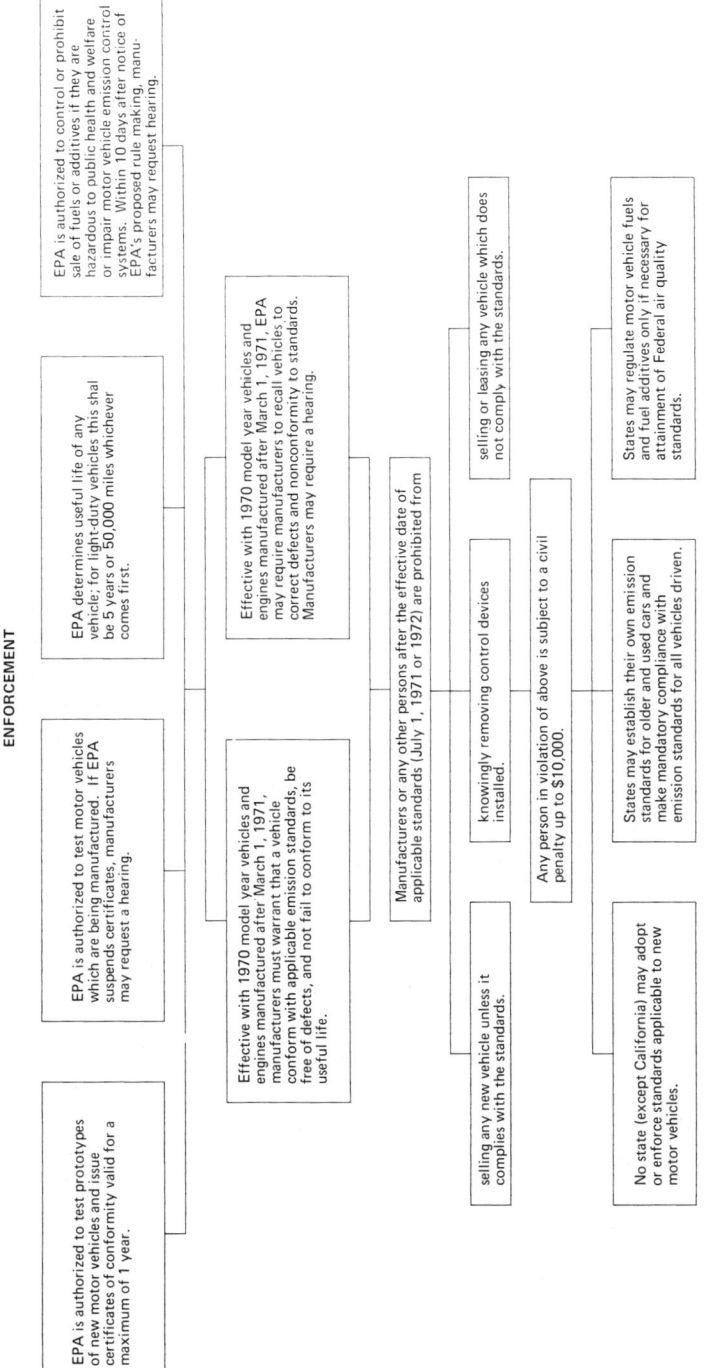

- EPA is authorized to test prototypes of new motor vehicles and issue certificates of conformity valid for a maximum of 1 year.

- EPA is authorized to test motor vehicles which are being manufactured. If EPA suspends certificates, manufacturers may request a hearing.

  - Effective with 1970 model year vehicles and engines manufactured after March 1, 1971, manufacturers must warrant that a vehicle conform with applicable emission standards, be free of defects, and not fail to conform to its useful life.

    - selling any new vehicle unless it complies with the standards.

    - No state (except California) may adopt or enforce standards applicable to new motor vehicles.

- EPA determines useful life of any vehicle; for light-duty vehicles this shall be 5 years or 50,000 miles whichever comes first.

  - Effective with 1970 model year vehicles and engines manufactured after March 1, 1971, EPA may require manufacturers to recall vehicles to correct defects and nonconformity to standards. Manufacturers may require a hearing.

    - Manufacturers or any other persons after the effective date of applicable standards (July 1, 1971 or 1972) are prohibited from

      - knowingly removing control devices installed.

      - Any person in violation of above is subject to a civil penalty up to $10,000.

      - States may establish their own emission standards for older and used cars and make mandatory compliance with emission standards for all vehicles driven.

    - selling or leasing any vehicle which does not comply with the standards.

    - States may regulate motor vehicle fuels and fuel additives only if necessary for attainment of Federal air quality standards.

- EPA is authorized to control or prohibit sale of fuels or additives if they are hazardous to public health and welfare or impair motor vehicle emission control systems. Within 10 days after notice of EPA's proposed rule making, manufacturers may request hearing.

the standards," is a loophole easily available to persons tried by injunctive court action. A misdemeanor prosecution would be better, because it need not consider ignorance or intent, but only the fact of a violation. In order to prevent token fines given in the past, it would be appropriate to set not only maximum but also minimum fines for certain categories of violations.

A governor may request from EPA an extension to implementation deadlines, and the president may exempt any emission source from compliance if he determines it to be in the paramount interest of the United States. It is, however, not easily appreciable what could be of greater interest to the United States than the protection of the public health and welfare of its own people.

Critique regarding mobile sources. The Federal law controls only emissions from new vehicles. In order to be effective emission control should be coupled with the regulation of fuels and fuel additives, maintenance supervision, and other methods of control such as land-use and rapid transit systems.

Despite the fact that in the past the automobile manufacturers have conspired against the development of effective emission control devices for vehicles, the new legislation stipulates that upon request further extensions of emission control deadlines might be allowed (see Figure 6.2). It is certain that the requests will be made. EPA may grant an extension, if it "is essential to public interest or public health and welfare." This sounds rather odd, because what could be more in the interest of public health and welfare than the earliest possible elimination of vehicle pollution? Granting further extension to the deadlines of vehicle emission control will definitely jeopardize efforts to achieve locally acceptable air quality levels.

Federal law forbids the sale or import of cars without the proper control devices, but not the actual driving of such vehicles. Legislation pertaining to mobile sources, in the same fashion as that for stationary sources, provides loopholes such as the word "knowingly" which can lead to acquittal under injunctive court action. No distinction should be made between the violation of a traffic rule and vehicle emission standards. Not only maximum but also minimum fines should be specified for certain categories of violations. The proper maintenance of emission control devices is the responsibility of the purchaser or operator of a vehicle. State laws should force the individual to maintain his vehicle properly.

EPA has no authority to test vehicles already in use. State laws should regulate inspection of emission control devices. EPA has not given any deadlines on restrictions for fuel additives. The public and Congress should request from EPA progress reports of fuel control. Additionally, no deadline has been given on the enforcement of regulations to control pollution emission from aircraft. The contribution of pollution from aircraft will therefore require the public's special vigilance.

LEGAL ENFORCEMENT

The Clean Air Amendments of 1970 emphasize compliance with the act rather than punishment under the act [5]. EPA may bring civil action against a violator, which is preferable to criminal action, because civil procedures only require a preponderance of evidence, whereas criminal procedures require proof beyond reasonable doubt. State and local air pollution control ordinances enforce pollution violations either by misdemeanor or injunctive procedures. Two examples, the misdemeanor procedures of the Los Angeles County Air Pollution Control District, and the injunctive procedures of the Bay Area Air Pollution Control District (San Francisco), will demonstrate the differences [9].

Misdemeanor enforcement. The Los Angeles county ordinances are based on a permit system which screens all plans prior to the actual construction of a potential pollution source. After the construction, operating permits are issued if the equipment performs according to prescribed specifications. Violation of the ordinances is a misdemeanor punishable by fine and/or imprisonment. Between 1955 and 1965 $710,000 in fines were collected from about 29,000 convictions. About 96 percent of all prosecutions led to convictions, because in misdemeanor proceedings neither ignorance of the law nor action in good faith are considered valid excuses. The accidental release of hazardous pollutants is considered no less an offense than an automobile accident.

Injunctive enforcement. The Bay Area Air Pollution Control District ordinances are based on performance standards only. In contrast to Los Angeles, the Bay Area's ordinances neither specify the design of air pollution control equipment and the construction characteristics of a potential pollution source, nor do they require permits for construction and operation. If performance standards are violated, a hearing board will first take abatement procedures; only in the second phase will a court take injunctive action.

Both the misdemeanor and the injunctive enforcements have some advantages and disadvantages. A major advantage of injunctive enforcement is that facts supporting a complaint have only to be probably true, whereas misdemeanor prosecution requires the establishment of facts beyond a reasonable doubt. Injunction is more directed toward future compliance, while misdemeanor punishes past offenses. Although injunctive cases are usually—on agreement of both parties—tried by a judge without a jury, an adjudication is difficult to appeal, because violation of an injunction is an act in contempt of court. A major disadvantage of injunctive enforcement is that ignorance of the law has to be considered and intent of the violator has to be proven, whereas misdemeanor enforcement need not consider ignorance or intent but just the fact of violation. It is clear that the Federal Clean Air Amendments of 1970 provide a considerable number of loopholes for air quality violators, because their intent to pollute and ignorance of the law has to be proven (see "knowingly" and "good faith effort" in Figures 6.1 and 6.2). It would appear that Los Angeles county—perhaps because of its aggravated air quality conditions—has the strictest enforcement procedures to control air pollution.

An example of court action. The following is an actual appeal case involving the Air Pollution Commission of Pennsylvania, the appellee, and a coated materials company, the appellant. Hoping to appeal a previous adjudication against them, the company put forth the familiar objection that the state agency's findings were not sufficiently substantiated by facts and conclusions of law [10]. The appeal came before the Court of Common Pleas of Dauphin county, Pennsylvania, in October, 1969. The court's "Opinion" gives the following background information on the case:

The coated materials plant is located in a valley; homes on a slightly higher elevation border the plant. Coil coating is the sole operation of the plant, which throughout its paint baking and water cooling processes emits odors and gases.

The chief witness for the Air Pollution Commission, an air pollution control engineer, testified that the odors from the plant were acrid and pungent and a source of nuisance and irritation to residents in its vicinity. Another witness, the chairman of the Regional Control Association, testified that since 1961 he had received 47 individual complaints from property owners in the vicinity of the plant.

The coated materials plant, or the appellant, admitted that by the nature of its operation odors are present within the plant. But it contended that the evidence that odors were present outside the plant was "vague, unconvincing and in some instances incredible." As an example the company cited the testimony of a female property owner whose husband had apparently been fired from his job at the plant. The company's appeal was thus actually based on the credibility of the evidence. The company additionally emphasized present extreme financial difficulties bordering on insolvency.

The court on reviewing the appeal as to the sufficiency of evidence, quoted from another case: ". . . it is the duty of the court to determine whether findings. . . are supported by the substantial and legally credible evidence required by the statute and whether the conclusions deduced therefrom are reasonable and not capricious . . . ." The court ruled that without doubt the evidence was substantial, and that the witnesses were certainly not unreasonable or illogical or conflicting in their statements.

In May 1970 the appeal of the coating plant from the adjudication by the Air Pollution Commission of October 1969 was dismissed. The motion to quash was denied, thus sustaining the adjudication which ordered the company to (1) reduce the emission of odorous air contaminants from its plant to such a level that they are no longer detectable beyond the plant's property line on or before June 1, 1970; and (2) submit a plan to the Department of Health which gives a detailed description of the control devices to be used, monthly progress reports, and a schedule of the expected completion date of each phase of the control plan.

SUMMARY

The major purposes of any air pollution legislation are (1) to protect the health and welfare of man, (2) to preserve plant and animal life, (3) to prevent any damage to physical property, and (4) to guarantee an aesthetically pleasing environment. The enforcement of air pollution legislation requires constant public awareness and public supervision.

Air pollution legislation dates as far back as 1306, when during the reign of Edward I of England violators of the pollution law were even executed. Air pollution in the United States is Federally controlled by the 1967 Air Quality Act as amended by the 1970 Clean Air Bill. Federal standard setting and implementation are controlled by the Air Pollution Control Office within the newly established Environmental Protection Agency. States and local political subdivisions may, however, set their own more stringent controls. The components of the 1970 Clean Air Amendments are discussed. The major changes and a critique of the new law are presented.

The air pollution control ordinances of the Los Angeles County Pollution Control District and the San Francisco Bay Area Air Pollution Control District serve as examples to explain misdemeanor and injunctive control procedures. An actual case of court action in Pennsylvania is reviewed.

Urban Planning and Air Pollution Control

Urban development and atmospheric pollution are directly related to population increase. Unfortunately, the people who are mainly responsible for an urban development, namely mayors, local legislators, businessmen, real estate speculators, and city planners, in most cases show little concern about the consequences which might ensue from their profit-orientated planning.

Effective air pollution control through city and regional planning would include the following considerations: An emission inventory would first identify the major air pollution sources and their pollutants. An air pollution sampling network would then delineate polluted from cleaner areas. The economic and social development of an urban area and its dependence on pollution sources would have to be assessed. However, insufficient control techniques and economic considerations often prevent effective air pollution control. This fact enhances the importance of the city and regional planner in controlling air pollution. Through proper application of his tools such as zoning, site selection, the use of green areas and buffer zones, street, highway, and rapid transit design, slum clearance and urban renewal, the planner can decisively contribute to air, water, waste, and noise pollution control. In the following paragraphs it is implied, though not always stressed, that the city planner through his actions can either adversely or positively change the local weather and climate; that he can provide comfort and amentiy for the citizens; and that he can ruin or improve the aesthetics of the original landscape.

GREEN AREAS AND AIR POLLUTION CONTROL

The preservation of green areas in cities is one major planning measure by which city planners and landscape architects can influence the climate and air hygienic conditions of an urbanized area. Green areas in an urbanized region comprise all open spaces from parks, woods, and cemeteries to recreation grounds, gardens, and bare land. The typical land use pattern for a selection of large United States cities shows that on the average about 30 percent of the total urban area is used for residential purposes [11]. Another 26 percent is unbuilt-up open ground. Streets and highways cover more than 23 percent of the city area; and industrial, commercial, and public institutions use up the remaining 20 percent.

Aerosols. It has long been suspected that green areas can filter out dust, soot, and fly ash from the atmosphere. In order to prove this, experiments are usually designed to record pollution levels inside and outside of green areas.

For Hyde Park, a recreation area of only one square mile in size in the center of London, an average reduction in the smoke concentration of 27 percent was found [12]. The cleansing effect of a park-like cemetery on the concentration of particulate pollution in an industrial valley in Cincinnati is shown in Figure 6.3 [11]. Under a light easterly airflow the particulate emissions from industrial plants and a city incinerator diffuse westward through the green area. It is important to note that over a distance of less than 200 yards the vegetation is able to reduce the concentration of suspended particulates by about 120 $\mu g/m^3$.

In a detailed study in the United States it was found that a dense coniferous forest can reduce the concentration of ragweed pollen by 80 percent [13]. Deciduous trees were much less effective. Even the small Aitken nuclei (0.1 μm) showed a reduction of 34 percent in a coniferous forest, and 19 percent in the deciduous forest. Although coniferous trees are more effective in filtering out pollutants than deciduous trees, coniferous plants should not be used indiscriminately for filtering purposes, because they are more susceptible to pollution damage.

An interesting study in the Soviet Union was set up to find out which tree genera exert the greatest filter effects. For this purpose the dust per unit area of leaf surface was weighed. The best vegetative dust filters in descending order and in g/m^2 of leaf surface were lilac with 2.33; maple with 1.11; linden with 0.61; and poplar with 0.26. It was calculated that 400 poplars which are the poorest dust collectors, spread over 2.5 acres, would filter out 0.375 tons of dust during the leaf-bearing season.

Most recent studies carried out along highways in the United States show that lead concentration falls off quickly with distance from the highway [14]. Plants and soils along these highways showed very high lead concentrations which were apparently obtained through both leaves and roots.

Gases. With the increasing automobile traffic the filtering effect of green areas on gaseous pollutants from busy highways would be of particular interest. Carbon monoxide levels were studied in a residential area free from industrial pollution. The houses were separated from the street by a dense stand of trees and brush 10 meters wide and about 6 meters high. A 1-9 mph wind was blowing from the street toward the houses throughout the experiment. At the sidewalk, 22 meters away from the road, carbon monoxide was reduced by 44 percent. About 30 meters away from the street, at a balcony on the fourth floor, carbon monoxide was even reduced by 54 percent. In another Russian study it was found that a park in

118　ATMOSPHERIC POLLUTION

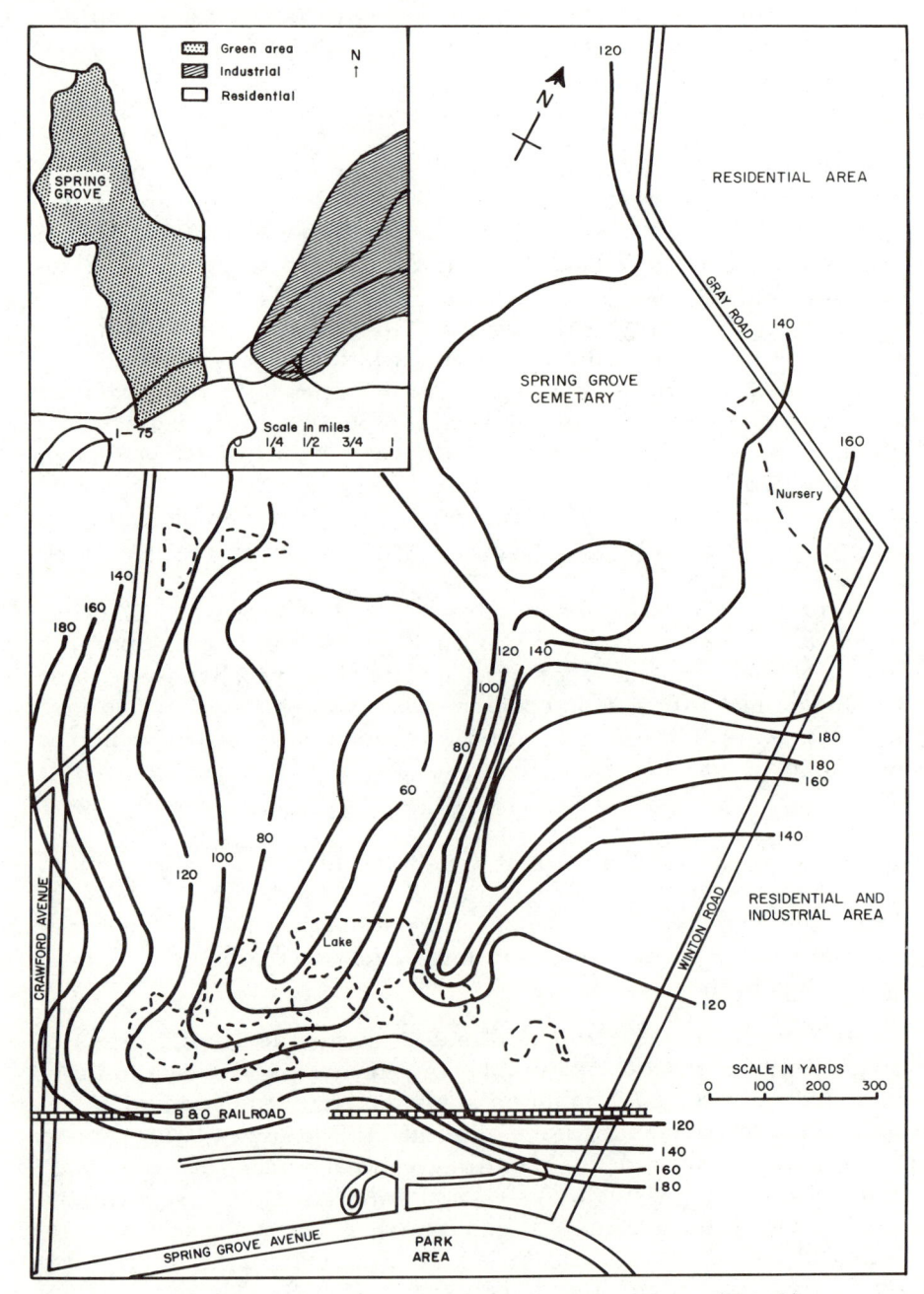

Figure 6.3　The effect of a green area on mass concentration ($\mu g/m^3$) patterns in an industrial valley, 8:00-8:45 A.M. E.S.T., 7/14/70

Leningrad reduced the sulphur dioxide concentration by about 50 percent as compared to an adjacent open area.

URBAN RENEWAL AND AIR POLLUTION CONTROL

The rural population is continuously decreasing while the urban population is constantly increasing. The end result is overcrowding in cities, slum conditions, an increasing air, water, and waste pollution problem, high taxes for poor services, and an increasing crime rate in a rootless society [15]. Today all cities of 50,000 and over in the United States have some kind of a pollution and noise problem [16]. Since more than 70 percent of the total population now lives in urban areas, this means that presently some 140 million people are the victims of industrialization, urbanization, and mechanization in our society.

Programs of urban renewal. Urban renewal as a planning measure has been extensively used to control atmospheric pollution. There are four major programs for renewing a city [15].

1. Rehabilitation of existing housing. If the overall structure is in satisfactory condition, houses are modernized and equipped with pollution-free appliances such as central heating and gas or electric heating systems.
2. Demolition of slum areas. Selected city quarters are demolished and rebuilt at the same or at other sites. More attention is given to recreational, educational, transportation, and hygienic requirements.
3. Construction of new housing and shopping centers at the periphery of cities. This perpetuates the urban sprawl and creates new transportation and hence pollution problems.
4. Construction of new towns outside the city limits of existing towns. In this case the latest state of the art in traffic planning, location of residential, commercial, administrative, and industrial complexes, and recreational layout can be applied.

The new town of Columbia, midway between Washington, D.C., and Baltimore, and planned to absorb the outmigration from these cities, is presently being built taking climatic, air hygienic, noise, traffic, economic, and amenity factors into consideration. Columbia is planned to absorb a population of 180,000. It has been estimated that from 1975 onwards, a city the size of Columbia will have to be built every year in the Boswash area to absorb the normal population growth. It is clear that even the most ingenious urban planning is doomed to failure, if population growth cannot be brought under control.

Modern residential planning. A few examples are now given for a better residential planning. The sketches in Figures 6.4 to 6.6 are based on new towns constructed by architect and town planner Reichow in Germany. Since new residential quarters are usually

120 ATMOSPHERIC POLLUTION

constructed away from polluting industry, it is the automobile which provides the major nuisance and harm from pollution and noise. Thoughtful planning as shown in Figure 6.4 takes the potential health hazard into consideration by shielding sidewalks and living quarters from automotive pollution and noise through earthwalls and brush.

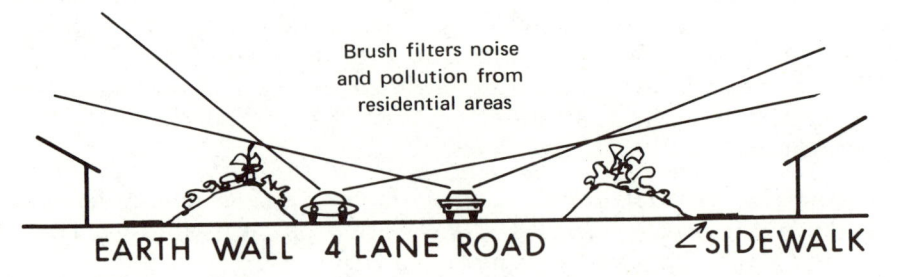

Figure 6.4 Road design in residential area.

Proper residential design would also avoid major through-roads. The no-outlet circles with one-family housing in a park-like landscape as shown in Figure 6.5 are planned with natural barriers against overcrowding, pollution, and noise. The argument that lack of space

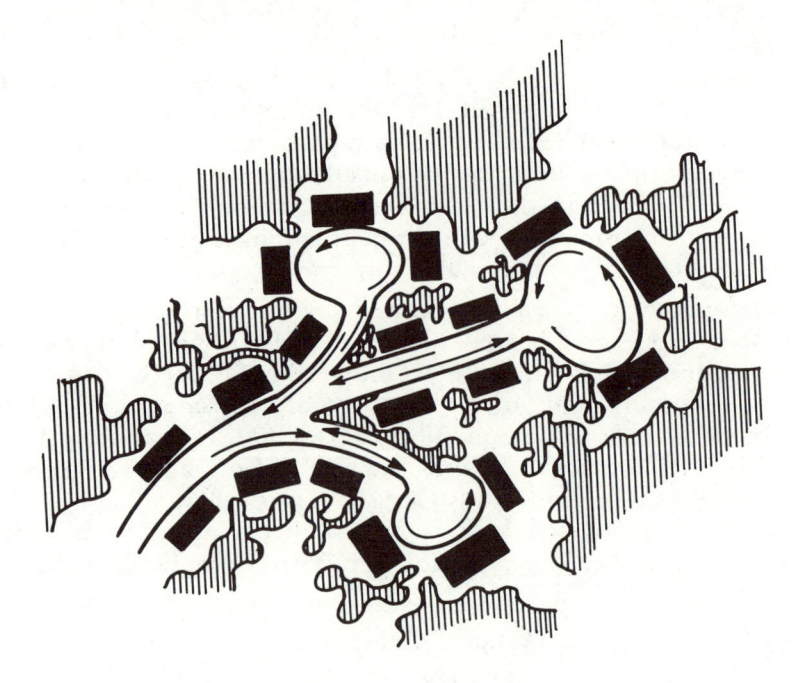

Figure 6.5 Residential design.

and housing shortage would not permit such spaceous planning is deceptive. Europe has much less space than the United States and can still afford such planning; furthermore, the only cure for lack of space and overcrowding is population control.

In purely residential areas automobile parking is clearly the greatest source of pollution and noise. On cold and calm winter mornings when turbulence and dispersion is almost zero, a single automobile warming up in neutral gear could shroud a whole neighborhood in toxic fumes. Also, the noise from people coming home from a late party and slamming car doors could be avoided if the parking lots were kept separate from the residential quarters. A parking lot covered with trees as shown in Figure 6.6 would not only preserve the characteristics of a park-like landscape, but would also prevent cars from overheating and excessive fuel loss due to evaporation on hot summer days. If garages are built on the property, they should at least be detached from the house. This again would avoid the constant stale gasoline stink and chronic gas intoxication typical of drive-in garages.

Figure 6.6 Parking lot.

TRAFFIC SANITATION

Automobiles, pollution, and noise. Closely related to urban renewal problems are those of traffic sanitation. As the urban area grows, so does the number of vehicles, which in turn means more roads and more pollution and more noise. In a city, three levels of pollution have to be distinguished:

1. The background concentration of the urban environment,

which is typically recorded by the pollution networks, and which is indicative of long-term chronic health effects.

2. The sidewalk and in-traffic concentrations, which are superimposed on the urban background pollution levels. Short-term pollution concentrations measured on busy streets represent peak exposures and are responsible for acute adverse health effects. During rush-hour conditions we recorded at breathing level on Kalakaua and Kapiolani Avenues in Waikiki, Honolulu, peak concentrations of 288 $\mu g/m^3$ for suspended particulates, and 10 ppm for carbon monoxide. The urban background readings at the University of Hawaii for suspended particulates were about 15 $\mu g/m^3$ and near zero for carbon monoxide. The peak levels are comparable to those found in other United States cities of equal size.

3. The indoor pollution levels, which are strongly related to outdoor pollution concentrations. A comparison of carbon monoxide levels indoors with those outdoors at 55 feet from the sidewalk at 54th Street and Sutton Place in New York showed that on a weekday the indoor pollution remains slightly above the outdoor level over the whole 24-hour period [17]. The carbon monoxide oscillated between a morning minimum of 5 ppm and rush hour peaks of about 15 ppm.

Street and higway design. Streets and highway systems are an integrated part of the urban complex and should consequently be treated as such in any urban renewal. From an air pollution and noise point of view this means that stop-and-go traffic should be avoided, particularly near residential and shopping areas. At moderately dense traffic and with courteous drivers, traffic circles would keep the traffic moving and avoid the high pollution concentrations during idling, de- and acceleration. The center of the traffic circle could embellish the road system with flower beds.

Highways, if they are to serve their primary function which is to provide fast communication arteries, should have cross-free exits and entrances, over- and underpasses, which keep the traffic moving (Figure 6.7). Economic factors, right-of-way, solid design incorporating safety and aesthetic factors, reduction of air and noise pollution are all paramount in a good highway design. The following individual aspects should be additionally considered:

1. Aerometric and diffusion studies should determine the safe distance between a highway and residential areas.
2. A safe width for individual lanes, the maximum permitted speed, and the size of vehicles should be worked out.
3. Dividing strips between opposing lanes should be wide enough and have enough vegetation to avoid blinding from oncoming headlights.

MEASURES OF AIR POLLUTION CONTROL 123

4. All heavily-traveled roads should have elevated ramps or vented underground walkways for pedestrians.
5. Tunnels should only be employed if the ventilation system can keep the concentrations below the highest outside pollution levels.
6. For dispersion reasons no recessed highways should be constructed in downtown areas.

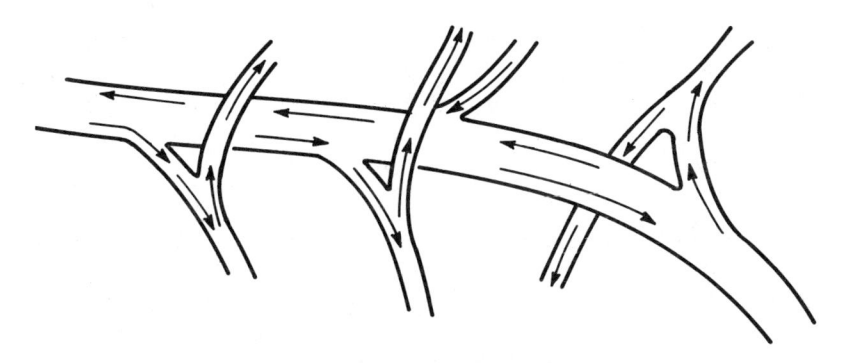

Figure 6.7 Highway design.

Rapid transit systems. Since the number of automobiles increases even faster than the population, it is conceivable that in the near future traffic by the most common means of transportation, the automobile, will come to a standstill. In most downtown areas traffic by automobiles in the 1960s was already slower than traffic by horse and carriage in the late nineteenth century. It is therefore high time to provide other kinds of transportation.

The term "rapid transit system" is as often misunderstood as used. Diesel buses running on the same roads as all the other automobiles are no rapid transit system, nor are tramways if their tracks can be blocked by other traffic.

Rapid transit systems using their own tracks either above or underground exist in many large cities. Subways, although more expensive to construct, are preferable because they produce no noise above ground and do not dissect the city with tracks. Subways in London and Montreal are safe, clean, and on time; subways in New York, Philadelphia, and Chicago are obsolete, dirty, unsafe, and unpunctual. A new management might make these subways again attractive for the public.

Westinghouse has been experimenting with a transit expressway in South Pittsburgh. The system runs quietly on rubber wheels; it is pollution-free, propelled by two 60-horsepower electric traction motors for each vehicle at a speed of about 50 miles per hour. This

124   ATMOSPHERIC POLLUTION

mass transit demonstration project, 9,360 feet long, cost about $5 million. For comparison, the anticipated San Francisco Bay Region rapid transit system will cost about $925 million for a 75-mile network [18].

Monorails are other variations of rapid transit systems already successfully in use in Tokyo. They can be of the hanging type, the predecessor of which has been running in Wuppertal-Elberfeld, Germany, for over half a century, or they can be of the Tokyo-type (Figure 6.8). The computerized one-cabin, one-person transit system (Figure 6.9) has been suggested particularly for the United States cities. This idea seems to be unfeasible, however, because it would presumably produce on tracks the same congestion experienced on roads. A large number of people can only be transported quickly in large compartments.

Figure 6.8 Types of monorails.

CONTROL OF INDUSTRIAL POLLUTION

In the past, availability of labor force, closeness to raw materials, transportation facilities, markets, and water supply were the decisive criteria for the selection of an industrial plant. Neglect of climatic, dispersion, and air hygienic factors at the planning stage subsequently proved to be a costly affair, if the firm had to pay for damaged crops or the depreciation of property in the vicinity of plants. It is always cheaper to design control devices into a new plant than to add such devices later. In many cases firms have threatened to move their plants elsewhere if forced to add control devices. Since the designation of Air Quality Control Regions all over the United States with air pollution standards for all major pollutants in 1971, this has become an empty threat.

Because it is not at present technologically possible to completely eliminate all major pollutants at the emission source, an intelligent assessment of the meteorologic, topographic, and potential adverse effects on land use and the population should be carried on prior to the construction of any plant. The following example, for which some of the data are taken from Katz [19], will show such an assessment at the planning stage.

Site selection. A new coal burning power plant is to be built on the shore of Beaver Lake which is 40 miles long and 15 miles wide located at 40° N in the central United States (Figure 6.10). The plant will burn about 5,000 tons of pulverized coal daily, which, at a sulphur content of 3 percent, would emit about 300 tons of sulphur dioxide and a large amount of fly ash. The company intends to install electrostatic precipitators with a dust collecting efficiency of 95 percent, but the plant may still emit more than 25 tons of fly ash per day. With a proposed effective stack height (physical height of the stack plus the plume rise) of 400 feet and an average wind speed of 10 mph at the site, most of the ash deposition will occur within about 1600 feet of the plant. If we suppose that 75 percent of the ash is deposited within the radius of one mile, i.e., over an area of more than three square miles, then a monthly fly ash rate of about 54 tons per square mile can be expected. Naturally, the sector with the prevailing wind direction can expect a greater deposition rate. It is suggested that the plant be located on about 2,500 acres of its own land to avoid damage to adjacent private property.

The terrain in the vicinity of the lake is quite flat. The coal will be transported to the site by steamers using the Black River. The major population centers in the vicinity of the lake are Urbana (300,000), Laketown (200,000), and Pinopolis (100,000). The meteorological station for the period 1961-1970 reported an average

126 ATMOSPHERIC POLLUTION

Figure 6.9 Individualized transit system.

wind speed of 10 mph with a 65 percent frequency from the southwest. The frequency of inversions below 500 feet per annum was 25 percent.

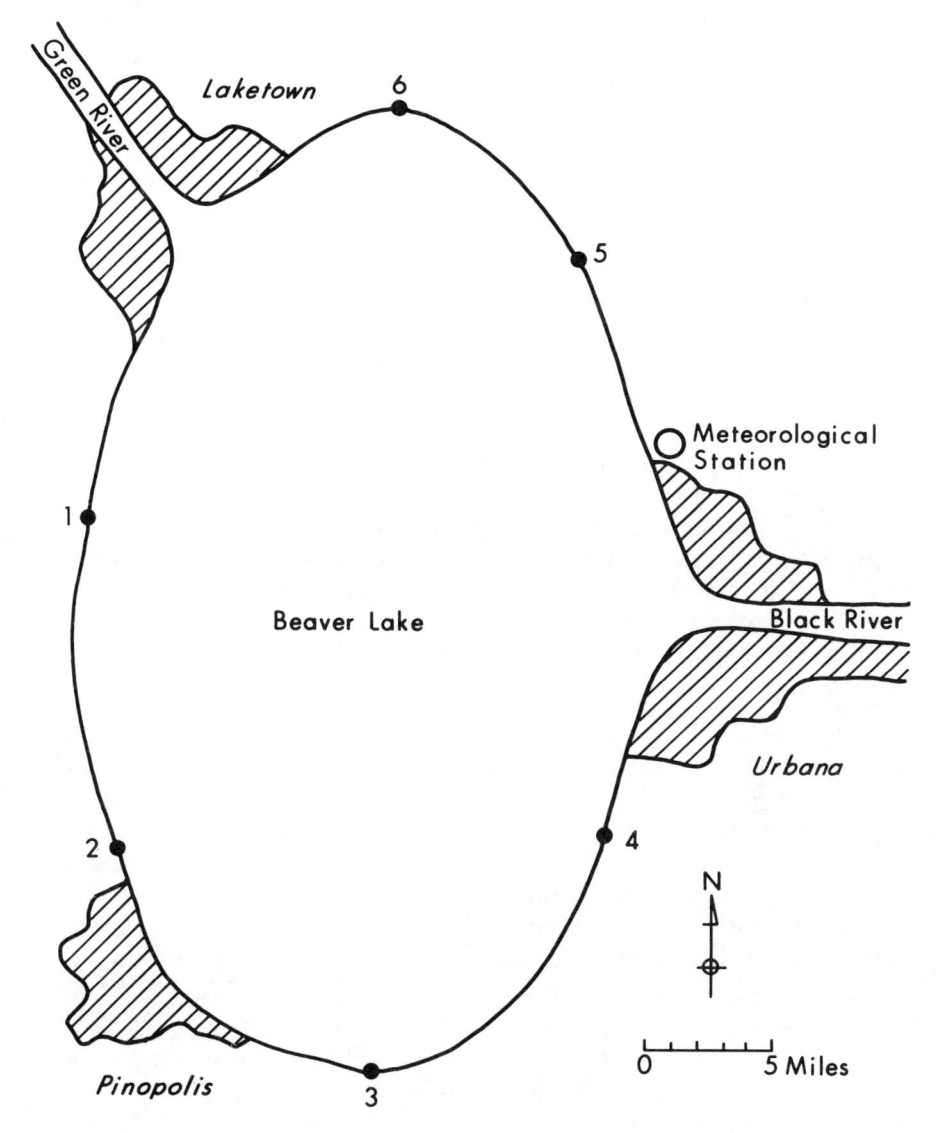

Figure 6.10 Site selection for a power plant.

The six locations shown in Figure 6.10 were proposed by the power company as possible sites. Experts were called in to determine the best possible site taking all the above factors and information into consideration. It was immediately clear that the provided meteorological information was too scanty to base any sound judgment on it. Thus, over a period of three months the following meteorological and diffusion data were collected: the frequency of wind speed and direction at the proposed sites; atmospheric stability and the frequency and depth of inversions (using radiosondes, tethered balloons); wind variation with height (using pibals); tracer studies (using bivanes and rotorod samplers); also the frequency and duration of anticyclonic (high pressure) weather with stagnating air masses; the lake and land breeze effect; and the flood height around the lake.

Additionally, a grading system for all six sites was prepared judging the terrain and vegetation, the population density, the availability of manpower, distance and cost of fuel transportation, cost of power transmission, cost of land, and other existing sources of pollution in the vicinity. Finally, site number 5 was recommended as the best location for the following reasons: (1) It is best from a wind direction point of view. (2) It is far enough away from North Urbana not to present a pollution hazard. (3) It is closest to the fuel supply and the labor market. (4) Under stagnating air masses with shallow inversions the whole area could be affected by the emissions. (5) It is therefore recommended to raise the effective stack heights to 600 feet, which would prevent inadmissible surface concentrations. (6) It is recommended to update the air pollution control equipment continuously in order to reduce global pollution concentrations with their implication on climatic changes.

Zoning. The chief tool of the urban planner to control, limit, and eliminate air and noise pollution is zoning [20, 21]. In residential areas zoning usually regulates the use of the land. For pollution purposes one is concerned with zoning the use of the atmosphere above. Zoning may be for the location of industrial or power plants. In this case the planner should evaluate real estate values, population and economic growth trends, the topography and the weather conditions, other pollution sources, and the concentrations and possible adverse health effects at receptors. Residential zoning usually regulates the lot size and the arrangement and height of buildings on the property. In connection with air pollution control it is important to supervise what type of industry will be permitted close to residential areas, and what will be the potential pollutant concentrations from such industrial sources.

Performance standards. Atmospheric pollution from industrial sources can finally be controlled by setting performance standards [22]. There are certain times of the day when urban dispersion is either absent or turbulence is so strong that pollutants are carried to the ground rather than dispersed upwards. These unfavorable conditions last from sunset to about 1-2 hours after sunrise. During these times pollutant emissions should be drastically reduced. Under adverse dispersion conditions a company could be asked or forced by law to change to a fuel with a lower sulphur content and less volatile matter.

SUMMARY

The decision on urban development rests largely in the hands of mayors, local legislators, businessmen, and real estate speculators, who either show little concern or have little understanding for the complex problems accompanying the modification of a natural landscape. It is clear that urban and regional planning, or rather the absence of it, has not only aggravated but also produced some of the problems related to air, water, waste, and noise pollution. Through thoughtful application of such methods as zoning, site selection, highway and rapid transit design, slum clearance and urban renewal, and the thorough understanding of the climatic and air hygienic factors, the urban and regional planner can play a decisive role in securing a livable urban environment.

Climatological factors such as solar radiation, greenhouse and heat-island effect, ventilation, and precipitation markedly influence the urban air hygienic situation. By integrating these factors in his planning scheme the urban planner can bring them to beneficial use for the general public. In this respect green areas have been widely used to modify the local climate and to filter out air and noise pollution.

Various programs of urban renewal, such as rehabilitation of existing housing, demolition of slum areas, and construction of new housing and town centers can help to control air pollution. Examples of modern residential planning, street and highway design, and rapid transit systems demonstrate a planner's potential in controlling air pollution. In the specific area of industrial pollution control well-advised developers and planners make use of proper site selection and zoning procedures, and adhere to the prescribed performance and emission standards.

Carefully controlled planning is a necessity. Planning will eventually have little impact if population growth cannot also be brought under control.

REFERENCES CITED

[1] H. C. Wohlers, et al., "Can Air Pollution Be Controlled by Legislation?" *Scientia* 104 (681-2), Series 8, 58-64, Jan./Feb., 1969.
[2] S. Edelman, "Air Pollution Control Legislation," in A. C. Stern, ed., *Air Pollution*, 2nd ed., vol. 3, Academic Press, Inc., New York, 1968, pp. 553-577.
[3] J. J. Schueneman, "Air Pollution Control Administration," in A. C. Stern, ed., *op. cit.*, pp. 719-796.
[4] Air Quality Act, Nov., 1967, Public Law 90-148.
[5] Clean Air Amendments, December 1970, Public Law 91-1783.
[6] *Air Conservation*, Report Air Conservation Commission of the AAAS, Publ. No. 80, Washington, D.C., 1965, pp. 212-233.
[7] USDHEW, PHS, *Guidelines for the Development of Air Quality Standards and Implementation Plans*, NAPCA, Washington, D.C., May, 1969.
[8] *Your Right to Clean Air, A Manual for Citizen Action*, The Conservation Foundation, Washington, D.C., Aug., 1970.
[9] Air Pollution Control, *Organization* (Community Action Guide No. 2), *Inspection and Enforcement* (Community Action Guide No. 4), National Association Counties Research Foundation, Washington, D.C.
[10] Court Decision of the Court of Common Pleas of Dauphin County, Pennsylvania, No. 778 Commonwealth Docket 1969, obtained from the Assistant Attorney General of the Commonwealth of Pennsylvania.
[11] W. Bach, "Seven Steps to Better Living on the Urban Heat Island," *Landscape Architecture*, 61(2), 136-138, 141, 1971.
[12] A. R. Meetham, et al., *Atmospheric Pollution, Its Origins and Prevention*, 3rd rev. ed., Pergamon Press, Inc., New York, 1964.
[13] H. Neuberger, et al., "Vegetation as aerosol filter," *Biometeorology 2, Prcdgs. 3rd Int. Congr.*, Pergamon Press, Inc., New York, 693-702, 1967.
[14] H. L. Motto, et al., "Lead in Soils and Plants: Its Relationship to Traffic Volume and Proximity to Highways," *Env. Science Technology* 4(3), 231-251, 1970.
[15] P. R. Achenbach, "The City: A Challenge to Engineering and Political Sciences," *A.S.H.R.A.E. Journal* (11)3, 33-38, Mar., 1969.
[16] F. W. Herring, "Effects of Air Pollution on Urban Planning and Development," presented at Nat. Conf. Air Poll., PHS, Dec. 10-12, 1962.
[17] J. C. Fensterstock, et al., "Reduction of Air Pollution Potential Through Environmental Planning," paper presented at A.P.C.A. meeting, St. Louis, 1970.
[18] "Air Pollution and Urban Development," chapter 7 in *Air Conservation*, Rpt. Air Cons. Comm. AAAS, Publ. No. 80, Washington, D.C., 1965.
[19] M. Katz, "City Planning, Industrial Plant Location and Air Pollution," in P. L. Magill, et al., eds., *Air Pollution Handbook*, McGraw-Hill Book Company, New York, 1956, Section 2.
[20] F. N. Frenkiel, "Atmospheric Pollution and Zoning in an Urban Area," *Scientific Monthly* 82(4), 194-203, Apr., 1956.
[21] P. M. Stern, "Planning and Zoning: Can They Be Made to Work for Clean Air?" presented at Nat. Conf. Air Poll., PHS, Dec. 10-12, 1962.
[22] W. J. Pelle, Jr., *Bibliography on the Planning Aspects of Air Pollution Control, Summary and Evaluation*, N. E. Illinois Planning Commission and USPHS, Dec., 1964.

CHAPTER 7

THE PUBLIC AND AIR POLLUTION CONTROL

Air pollution control can only be successfully achieved if the officials of a community have public support. Public support, however, can only be expected from well-informed citizens. Information and public support evolve from three separate phases: an educational program, a community relations program, and the action of public officials [1].

INFORMATION PROGRAMS

The major purpose of an educational program is to translate the complicated and highly technical problems related to air pollution control and implementation into a language the average intelligent citizen can understand. For example, if the complicated meteorological processes involved in interstate pollution transport and the present state of control technology are explained to the public, they will not request pristine air. But on the other hand, citizens should also be informed about all benefits from air pollution control in terms of either direct economic savings or indirect benefits by preventing morbidity and mortality caused by air pollution. It is important to create an informed citizenry as a counterbalance to the self-interests of the pollutors.

Mutual trust can best be developed if the various interest groups representing the pollutees and the pollutors openly discuss their problems and interests. Such a community relations program would, for example, explain what restrictions specific regulations impose on certain segments of the economy. It would also discuss whether the proposed regulations would adequately protect the general public. The desired quality of life should be discussed in all fairness by weighing the economic constraints against such intangibles as blue sky, good visibility, and a pleasing city environment. These discussions should make it quite clear that it is not the intention to impose on the pollutor unnecessary restrictions which might even put him out of business. However, it should be made equally clear that a priori nobody has the right to pollute the public good, air. In order to maintain a certain standard of living the general public may be prepared to put up with certain levels of pollution. If the public health and welfare is in danger, however, no compromise can be permitted.

Public officials, as the referees between the pollutors' and the public's interests, have the difficult task of deciding which of the desires of the different interest groups is best for the community. It is well-known that industry employs well-paid lobbyists, whose full-time job it is to pursuade legislators and other public officials to represent their interests. The informed and concerned citizen usually has little but his own time and keen interest in the well-being of his community to offer when he is educating public officials. This demanding role was called by the former chairman of New York City's Citizens for Clean Air, Inc., Hazel Henderson, "mature citizenship, the noblest responsibility in a democracy" [2].

PUBLIC AWARENESS OR APATHY

A number of public opinion polls have been conducted in order to find out how aware the general public is of air pollution problems, and how strong their willingness is to do something about it.

In 1964 a total of 1,002 people responded to a public attitude survey conducted in St. Louis city, St. Louis county, and Madison and St. Clair counties [3]. A good proportion of residents expressed some negative view on the quality of air in that region. They said that odor and smoke were the major pollutants bothering them, and they thought that factories and businesses were the greatest producers of these pollutants. More than 90 percent of the respondents suggested that some governmental agency should do more about air pollution control. As Table 7.1 shows, between 84 percent and 86 percent and 59 percent and 71 percent of the respondents in the St. Louis area indicated their willingness to spend $1 or $5 on air pollution control, respectively. The figures for the Charleston, West Virginia, area are quite similar.

A collection of opinions from nearly 3,000 people in Nashville, Tennessee, revealed the following results [4]: Awareness and concern about air pollution increased proportionately with air pollution levels. Also, women and people of a higher socioeconomic status showed a greater awareness and concern towards air pollution problems. Out of a population of 232,000, about 10,000 people considered air pollution a health hazard; about 50,000 were irritated by smog, and between 40,000 to 100,000 people were bothered by the soiling of surfaces and objects, decreased visibility, odors, and damage to property. About 12 percent of the respondents felt that the city officials did not do a good job of combating air pollution.

Table 7.1 Willingness to pay for air pollution control

Communities	Percent of people willing to pay						
	an additional $1			an additional $5		
	Yes	No	Don't know	Yes	No	Don't know
St. Louis county [3]	85.9	14.1		71.0	29.0	
St. Louis city	84.4	15.6		66.1	33.9	
Madison county	86.3	13.7		62.7	37.3	
St. Clair county	84.3	15.2		58.9	41.1	
Charleston [6]	82.3	13.5	4.3	64.5	27.0	8.5
South Charleston	91.6	7.9	0.5	83.7	16.3	no data
Nitro	91.3	2.0	6.7	No data		
Montgomery	93.1	4.9	2.0	80.4	16.7	2.9

Sources: J. Schusky, 1966 [3]; R. E. Rankin, 1969 [6].

Using face-to-face interviewing techniques, in 1959 a sample of 466 and in 1961 a sample 334 people was obtained in the Buffalo area [5]. Regarding perceptions of seriousness of community problems, it was shown that air pollution ranked in fifth place with 43 percent after unemployment (79 percent), juvenile delinquency (61 percent), car accidents (58 percent), and alcoholism (48 percent). The majority of people believed that air pollution adversely affects real estate values and that it is detrimental to health. People unanimously argued that air pollution control was a "good thing," but had no notion how it could be achieved. They had little confidence that something would be done, but in general they agreed that something should be done to control air pollution.

An interesting study relating to public interest and apathy concerning air pollution control programs was carried out for the Charleston, West Virginia, area [6]. About 65 percent of the respondents were aware of air pollution as a problem. The higher the pollution level the greater was the public's awareness. About 90 percent of the people seemed to be both aware and concerned. Thus, if public apathy existed, it would certainly not be due to a lack of awareness. People seemed to notice air pollution as a problem more in the community at large rather than in their own neighborhood. The informed state of the general public regarding air pollution problems was rather poor. Most people did not know what to do about air pollution. The general attitude ususally prevailed that the pollution conditions would remain the same. It seemed that this kind of public apathy was more related to what could and would be done rather than that there existed an air pollution problem.

From these interesting results, Rankin, a psychologist, draws the following significant and far-reaching conclusions [6] :

1. The arousal of anxiety or concern over a state of affairs, such as adverse effects of air pollution, is only effective if it is closely followed by action that reduces or eliminates the fear or the problem.
2. If no relief is in sight, people tend to defensively avoid all problems related to air pollution.
3. People would more likely follow recommended actions, if the level of fear elicited was low rather than maximal.
4. If people have no effective response to meet a threat, they either try to avoid the problem, or they develop a strong sense of personal invulnerability.

The arousal of an ecological conscience in the 1960s culminating in the environmental teach-ins and earth days in 1970 may have already reached its anticlimax in 1971. Opinion polls show that not Ralph Nader or John Gardner but rather John Wayne is again the hero for high school and college students. It is obvious that young people are disenchanted with the slow progress which is being made in solving our environmental problems. Solving our environmental problems is a long-term process. The problems will not be solved by dropping out and ignoring them, but rather by facing them and fighting a steady and dedicated battle.

ACTION PROGRAMS

Any action program thrives on the work that is being done in subcommittees or task forces. In connection with an air pollution action program, the Conservation Foundation [7] suggests the following committees:

1. *A study committee* whose function would be to evaluate air quality criteria, propose air quality standards, and develop implementation plans. The committee should preferably consist of experts in various fields, who can translate the highly technical problems into a language which the general public, legislators, and control officials can understand.

2. *A publicity committee* whose function would be to disseminate to the community the material which the study committee has prepared. The National Association of Counties Research Foundation [1] suggests further tasks. The publicity committee should publicize the establishment of citizens' groups, statements by public officials, a list of uncontrolled pollution sources and their effects, comparisons of adverse effects, air quality levels, and air quality standards (Table 7.2), meetings and public hearings, formation and activities of agencies, visits of experts, receipt of research and pollution control grants, periodic progress reports, and air pollution indices.

The publicity committee should also organize exhibits showing photographs, slides, and films, etc., of open burning at dumps, construction and demolition sites, and individual homes, smoke plumes from industrial plants and apartment buildings, fumes from motor vehicles, dust and fly ash settled on cars, homes with peeling and blackened paint, cities on clear and smog days, stacks with and without control devices, washing of buildings, vegetation damaged or killed by pollutants, acute pollution episodes, development of tumors and cancers in animals and man, maps showing pollution sources and amounts of pollutants, inspection trips by control officials, sampling equipment of control agency, and finally, activities of action groups such as meetings, demonstrations, and picketing.

3. *A legal action committee* whose function would be to appraise the air pollution legislation and work towards strict ambient air quality, performance and emission standards. The committee would help legislators to repeal legislation that is too lax, and draft and actively support stricter laws. It would supervise the strict enforcement of existing laws by reporting violators and instituting litigation. It would actively support officials with strong positions regarding environmental issues. In short, this committee will have a broad watchdog and whistle-blower function in the Nader sense.

4. *A communications committee* whose function would be to maintain the connections between the various action groups and the public, and organize workshops and meetings, inviting pollution control officials, city councilmen, state legislators, and officials of various viewpoints. This committee would also be responsible for obtaining all materials and publications issued by government agencies, industrial, commercial, private, scientific, health, conservation, and other organizations.

The distribution of the following information is of particular interest: Emission inventories which show the distribution of the major pollution sources and the amounts of certain pollutants; data from an aerometric network which include pollution and meteoro-

Table 7.2 Synopsis on air quality standards

Air Quality Criteria		Air Quality Level in Hawaii (1)	Adverse Effects (2)	Air Quality Standards						
				Federal		U.S. Regions (3)		Dept. of health			Hawaii 1971 UH task force
				primary	secondary	highest	lowest	Jan.	Apr.	adopted	
				Jan 30 (Apr. 30) 1971							
Suspended particulates ($\mu g/m^3$)	Geometric annual mean		Hazard to human health 80-100	75	60	Chattanooga 80	New York City 45				35
	Arithmetic annual mean	48	Material dammage 60-100					55	40-55	55	
	24-hour maximum*	96		260	150	Chicago 260	Los Angeles 100	100	80-100	100	60
Sulfur dioxide ($\mu g/m^3$)	Arithmetic annual mean	21	Plant injury 85	80	60	Cincinnati 57	Miami 9	25	25	20	10
	24-hour maximum*	.42	Hazard to human health 300	365	260	Chicago 486	Miami 29	114	80-114	80	25
	3-hour maximum*	105			1300	New York City 1430	Cincinnati 858	—	260	400	50

Table 7.2 Synopsis on air quality standards (continued)

Air Quality Criteria		Air Quality Level in Hawaii (1)	Adverse effects (2)	Air Quality Standards							
				Federal		U.S. Regions (3)		Hawaii 1971			
				primary Jan. 30	secondary (Apr. 30) 1971	highest	lowest	Jan.	Dept. of health Apr. adopted		UH task force
Carbon monoxide (mg/m^3)	8-hour maximum*	3.4	Time discrimination impaired		10	Chattanooga 12	New York City 9.2	9	3-5	5	3.5
	1-hour maximum*		12-17	15(40)					8-10	10	5 (4)
Hydrocarbons (μg/m^3)	3-hour maximum*	177	Photochemical hazard to human health 100	125 (160)	125 (160)	Dallas - Fort Worth 130	Springfield, Conn. 52	130	100	100	65
Photochemical oxidants (μg/m^3)	1-hour maximum*	57	Eye irritation, impairment of performance 60	125 (160)	125 (160)	New York City 118	Birmingham 40	120	80-100	100	40
Nitrogen dioxide (μg/m^3)	Arithmetic annual mean	47	Adverse effects 113	100 (100)					40-70	70	50 (4)
	24-hour maximum*	0-145	Acute bronchitis 118-156	250 (−)					100-170	150	100

Source: (1) State of Hawaii, 1969 [8], 1971 [9]; (2) USDHEW, 1969, 1970, 1971 [10]; (3) USDHEW, 1969 [11]; (4) proposed by chairman.

* Not to be exceeded more than once per year.

138   ATMOSPHERIC POLLUTION

logical observations, measurability of pollutants (Table 7.3), trends in pollution levels; information on topographical peculiarities which might affect air quality; surveys of adverse health effects; cost-effectiveness studies; existing and new air pollution control laws; and information on proposed and adopted air quality standards in other air quality control regions (Table 7.4). This kind of tabulation, which provides comparisons, is of value in setting the local air quality standards and in demonstrating the success of a public hearing.

Table 7.3 Measurability of air pollutants versus recommended air quality standards

Pollutant and determination method	Lower limit of determination	Accuracy	University of Hawaii Air Pollution Task Force recommended standard
Sulfur dioxide (pararosaniline method)	25 $\mu g/m^3$ (1)	4.6% (2)	Max. av. in any 24 hr: 25 $\mu g/m^3$
Suspended particles (high volume method)	1 $\mu g/m^3$ (3)	15% (4)	Max. av. in any 24 hr: 60 $\mu g/m^3$
Carbon monoxide (nondispersive infrared spectrometry)	no lower limit range: 0-48 mg/m^3	± 1%	Max. av. in any 8 hr: 3.5 mg/m^3
Oxidants (neutral buffered potassium iodide method)	20 $\mu g/m^3$ (5)	not defined (varies with oxidant measured)	Max av. in any 1 hr: 40 $\mu g/m^3$
Hydrocarbons (flame ionization method)	no lower limit range: 0-13,000 $\mu g/m^3$	± 1%	Max. av. in any 3 hr: 65 $\mu g/m^3$
Nitrogen dioxide (Jacobs/Hochheiser method)	20 $\mu g/m^3$ (6)	not available	Max. av. in any 24 hr: 100 $\mu g/m^3$

Notes: (1) For a sample of 30 liters. One can extrapolate to concentrations below 25 ug/m^3 by sampling larger volumes of air. (2) Relative standard deviation at 95% confidence level. (3) Sampler operated at av. flow rate 1.7 m^3/min for 24 hrs. (4) Relative standard deviation for average concentration of 39 $\mu g/m^3$. (5) At a sampling rate of 2 liters/min for 15 min. using 10 ml absorbing reagent. (6) At a sampling rate of 200 ml/min for 24 hrs. using 50 ml absorbing reagent.

Source: Federal Register, 1971 [12].

Table 7.4 Proposed and adopted air quality standards for sulfur oxides and particulate matter in a number of states

State	Particulate matter (micrograms per cubic meter: μg/m³)			Sulfur oxides (parts per million: ppm)		
	Before hearing		After hearing*	Before hearing		After hearing*
Indiana	annual	75	75	annual	.015	.015
Illinois	24 hr.	300	260	24 hr.	.20	.17
				1 hr.	.50	.42
Pennsylvania	annual	100	65	annual	.03	.02
	24 hr.	300	195	24 hr.	.25	.10
				1 hr.	.50	.25
New Jersey	annual	65	—	annual	.02	—
	24 hr.	195	—	24 hr.	.10	—
				1 hr.	.20	—
Delaware	annual	110	70	annual	.04	.03
	24 hr.	250	200	24 hr.	.22	.13
	(for urban areas)			1 hr.	.50	.30
Colorado	annual	55	—	annual	.009	—
	24 hr.	180	—	24 hr.	.05	—
				11 hr.	.50	—
Massachusetts	annual	80	—	annual	.031	—
	24 hr.	180	—	24 hr.	.105	—
				1 hr.	.280	—

(continued)

Table 7.4 (continued)

State	Particulate matter (micrograms per cubic meter: $\mu g/m^3$)			Sulfur oxides (parts per million: ppm)		
	Before hearing		After hearing*	Before hearing		After hearing*
Virginia	annual	80	60	annual	.05	.02
	24 hr.	125	100	24 hr.	.10	.10
Maryland	annual	65	—	annual	.01	—
	24 hr.	140	—	24 hr.	.05	—
				1 hr.	.10	—
				5 min.	.25	
District of Columbia	annual	65	—	annual	.02	—
	24 hr.	140	—	24 hr.	.08	—
				1 hr.	.25	—
Missouri	annual	75	—	annual	.02	—
	24 hr.	200	—	24 hr.	.10	—
Hawaii	annual	55	—	annual	.01	.007
	24 hr.	100	—	24 hr.	.04	.03

*dash (—) indicates no changes were made. In several cases, this reflects citizen acceptance or support of the state's numbers rather than any official neglect of citizen views.

Source: The Conservation Foundation, 1970 [7]. Supplemented for Hawaii.

All community action work—if successful—has to culminate in testimony presented at committee and public hearings. The following suggestions may be helpful when preparing a testimony. First and foremost the witness should be utterly clear in his own mind what he intends to get across. This does not mean that he necessarily has to be an expert in any of the fields dealing with air pollution. It does mean, however, that the speaker should add an important piece of information to the testimony. For example, if he observes that in polluted areas certain plants no longer grow, or hills become shrouded behind veils of smog, or an acquaintance develops a bronchial cough, then all these are significant observations.

The testimony would carry more weight if the witness can speak as the representative of an influential group, i.e., usually a large group of voters and tax payers. Finally, the testimony should be precise and simply worded. Arguments and quotations should be referenced and solidly documented.

SUMMARY

Effective air pollution control requires constant public awareness, public supervision, and public pressure. Only a well-informed citizenry can adequately fulfill this watchdog function. The information has to be conveyed in educational and community relations programs and the like.

Public opinion polls conducted in St. Louis and surrounding counties, Nashville, and the Buffalo and Charleston areas have shown the following interesting results: Almost all respondents suggested that governmental agencies should do more about air pollution control. A great majority of the people indicated their willingness to spend $1 or $5 per annum on air pollution control. Awareness and concern about air pollution seemed to increase proportional to air pollution concentrations. Women and people of higher socioeconomic status showed greater awareness and concern towards air pollution problems. People seemed to notice air pollution as a problem more in other communities than in their own neighborhood. Most people did not know how to abate air pollution. The arousal of anxiety or concern over adverse effects of air pollution is effective only if it is followed by action that reduces or eliminates the fear or the problem. If successful action fails to materialize, people either evade the problems or develop a strong sense of personal invulnerability.

The nature of various action programs is discussed and the work done in the different study, publicity, legal action, and communications committees is described. Community action usually culminates in testimony given before committees or at public hearings.

REFERENCES CITED

[1] Air Pollution Control, *Gaining Community Support*, (Community Action Guide No. 7), National Association Counties Research Foundation, Washington, D.C., no date.
[2] H. Henderson, "Citizen Action for Clean Air," *Congressional Record* 115(116), Washington, D.C., July 14, 1969.
[3] J. Schusky, "Public Awareness and Concern with Air Pollution in the St. Louis Metropolitan Area," *JAPCA* 16(2), 72-76, 1966.
[4] W. S. Smith, et al., "Public Reaction to Air Pollution in Nashville, Tennessee," *JAPCA* 14(10), 418-423, Oct., 1964.
[5] I. DeGroot, et al., "People and Air Pollution: A Study of Attitudes in Buffalo, N.Y.," *JAPCA* 16(5), 245-247, May, 1966.
[6] R. E. Rankin, "Air Pollution Control and Public Apathy," *JAPCA* 19(8), 565-569, Aug., 1969.
[7] The Conservation Foundation, *Your Right to Clean Air*, Washington, D.C., Aug., 1970, p. 44.
[8] "Aerometric Survey," Department of Health, State of Hawaii, 1969.
[9] "Proposed Ambient Air Quality Standards," Department of Health, State of Hawaii, Jan., 1971, rev. Feb., 1971, final version July 1971.
[10] "Air Quality Criteria, Summary and Conclusions," USDHEW, PHS, Feb., 1969, March, 1970, Jan., 1971.
[11] "Proposed or Adopted Air Quality Standards," USDHEW, PHS, *NAPCA*, Nov., 1969.
[12] "National Primary and Secondary Ambient Air Quality Standards, and Air pollution Control, "Federal Register 36(21) Pt. II, EPA, Washington, D.C., Jan. 30, 1971.

EPILOGUE: ATMOSPHERIC POLLUTION AND OUR FUTURE

In this generation many people are finally beginning to realize that our air is not an unlimited natural resource and that hazardous regional and even global air pollution levels already constitute an alarming reality. Some members of our society consider this situation to be one of the harbingers of inevitable ecological disaster. Others, however, cling tenaciously to the belief that a technological solution to the problem will soon be found. Both of these ideologies represent dangerous excuses for individual inaction. If one group of society says we are not doing anything because nothing can be done, and the rest of society says we are confident that someone else will soon find a solution to the problem, the end result might well be what Ian McHarg has expressed in his parable:

> The Atomic cataclysm has occurred. All life is extinguished save in one leaden silt where, long inured to radiation, persists a small colony of algae. They perceive that life exists in them alone and that two billion years must proceed for them to return to yesterday. They come to a single spontaneous, unanimous conclusion: Next time, no brains.[1]

Finding reasonable solutions to the complicated problems of air pollution, however, will require nothing less than the full mobilization of all of our available brain power. Representatives from the

[1] I. L. McHarg, *Environmental Improvement,* N.S. Dept., Agr. Grad. School, Washington, D.C., pp. 99-105, 1966.

physical, engineering, biological, medical, and social sciences must begin now to translate their research data and ideas into mutually understandable information which can then be used by environmental specialists working full-time toward finding practical solutions to the problem.

Future work concerning air pollution abatement must concentrate on education and guidance of the general public. Only well-informed people can possibly bring to bear the necessary public pressure that will cause the federal and local governments to promulgate meaningful legislation and cause administrative enforcement agencies to take appropriate actions against violators so that the quality of our air is indeed protected and enhanced.

Unless population growth can soon be brought under control, however, the possibility that we can maintain breathable air is nil. There is a vicious cycle that starts with more people, who demand more goods, which requires the conversion of more natural resources into more products, which unavoidably creates more pollution of our environment.

In order to break this vicious cycle we must develop a new philosophy of life. As individuals we must now begin to consider the ecological consequences of all of our actions. As a society of careless individuals we have recklessly degraded the quality of our air. As a society of environmentally aware individuals we are now faced with the consequences of our carelessness, and it remains to be seen whether we shall be equal to the monumental task of preserving our air.